GEOGRAPHIES OF URBAN SOUND

Geographies of Urban Sound

TORSTEN WISSMANN
Johannes Gutenberg-University, Mainz, Germany

Routledge
Taylor & Francis Group
LONDON AND NEW YORK

First published 2014 by Ashgate Publishing

Published 2016 by Routledge
2 Park Square, Milton Park, Abingdon, Oxfordshire OX14 4RN
711 Third Avenue, New York, NY 10017, USA

First issued in paperback 2016

Routledge is an imprint of the Taylor & Francis Group, an informa business

British Library Cataloguing in Publication Data
A catalogue record for this book is available from the British Library

The Library of Congress has cataloged the printed edition as follows:
Wissmann, Torsten.
Geographies of urban sound / by Torsten Wissmann.
pages cm
Includes bibliographical references and index.
ISBN 978-1-4094-6219-4 (hbk)
1. City noise–Europe–Case studies. 2. City noise–United States–Case studies. 3. City and town life–Europe–Case studies. 4. City and town life–United States–Case studies. I. Title.
TD893.5.E85W57 2014
363.74'1–dc23

2013045510

ISBN 13: 978-1-138-27014-5 (pbk)
ISBN 13: 978-1-4094-6219-4 (hbk)

Contents

List of Figures

List of Tables

About the Author

A native to Wiesbaden, Germany, Torsten Wissmann completed a M.A. (Magister Artium) in Geography, Philosophy and Pedagogy (2004) at Johannes Gutenberg-University Mainz, Germany, and a Dr. phil. (Ph.D.) in Geography (2009) at Johannes Gutenberg-University Mainz. His academic positions have been in geography departments at Johannes Gutenberg-University (2004–2007, Contract Teacher; 2008 to present, Assistant Professor) and the University of Texas at Austin (2011–2012, Research Fellow). Dr. Wissmann's research and teaching over the last five years has centered on issues of space, identity, sound and media. His key regions of research include Brazil, Malta, Germany, and the United States.

Acknowledgements

This book would not have been possible without the support of many people. I wish to express my gratitude to my mentor, Anton Escher, who always offered his help and enabled me to do my research in the United States. My deepest gratitude is also due to Leo Zonn, for his invaluable assistance, support and guidance, especially during my stay in Austin, TX. Special thanks go to all members of the departments of geography at both the University of Texas at Austin and the University of Mainz, who helped and inspired me in many ways, especially Stefan Zimmermann. I would also like to thank Thomas Bartsch (cartography), Aaron Negron (transcriptions), Tim Rizzo (book cover), Josh Rosenblatt (copyediting), and my student assistants Kathrin Hammes, Marc Hoffmann, and Theresa Schmitt for their continuous help and support. Not to forget Marc Boeckler and Christina Kennedy who encouraged the research project. My thanks also go to all interviewees for their openness. Finally, I also wish to express my love and gratitude to my beloved family, for their understanding and endless love.

Listen Up!

The cacophonous mix of sound that surrounds us in an urban environment is usually not disturbing because what we hear is an integral and accepted part of the urban dweller's life. Yet it is the visual element of the city that dominates the standard urban vocabulary of experience. Sound has long been neglected in urban studies. Traffic, music, language, and nature, as primary examples, help to create unique soundscapes essential to the place-based character of each city, and no consideration of these cityscapes should fail to include them. Substantive research on sound and soundscapes began over half a century ago as Acoustic Ecology with the World Soundscape Project, which focused on human-environmental relationships based on sound. There has not been a sustained body of research since then.

There are many ways to consider urban sound, from pure research to applied. As examples, raw urban sound can be recorded and analyzed for stress, tourist soundwalks can be developed to try to capture the auditive characteristics of the city for the visitor, or the urban soundscape can be deconstructed into single *sound events* that are subsequently put together to form artistic audio collages. Clearly, sound must be taken into account when considering a city's sense of a place, where a range of sounds – from music to human, from technology to nature – play integral roles in creating local identity and cultural boundaries. And, of course, governments at all scales, from local to federal, often become involved in issues of sound, from mitigation to promotion, as in the case of music. Taking into account both the urban soundscape and the impacts of sound on the urban dweller, I understand sound not as a by-product of urban life but as a fundamental part of urban life – something essential for understanding the city's sense of place.

The purpose of this book is to outline a framework for the study of sound within the urban landscape and to provide a unique look at the geographies of urban sound. Coming from a social geographic background, I am particularly interested in the effects of sound on the individual and the many ways sound influences how we engage the city as place, especially in terms of daily routines. I aim to uncover the socio-scientific potential of sound in the urban environment based on the understanding that sound cannot and must not be seen as detached from this urban landscape, but rather as a constituent element of the same. To my understanding, sound exists not only *within* the city; sound *is* the city.

I am convinced that studying the geographies of urban sound involves far more than geography as a discipline. Understanding the daily-lived experiences and environmental perceptions of the urban dweller and his/her *place* in the city should be of interest to all who wish to understand the city as home.

Chapter 1 presents the theoretical base of the book. Elements that comprise the individual's perception and the concept of a city's sense of place are described in detail. To this end, geographical studies on the sense of place and broader humanistic approaches to sound are explored, both within the context of urban settings. The chapter closes by addressing the many possible effects of sound on the lives of the urban dweller. Chapter 2 answers the call of the first chapter about the possible effects of sound. Findings are strongly informed by phenomenological thoughts derived from a literature based primarily in psychological, sociological, and medical works. Valuable input comes from acoustic design and architecture, ethnology and urban planning. Studies of the processes of hearing and the physiognomy of the human ear, the nature of wanted and unwanted sounds, and the absence of sound and the essence of silence form the heart of this chapter. Chapter 3 approaches empirical studies on urban sound with the methodology of acoustic ecology. Recoding, deconstruction, and classification of the soundscape are presented. Findings from three field projects in Portugal, England, and the United States adapt, translate, and evolve existing practice. The latest developments in mapping and acoustics are considered and utilized in these analyses of place and the perception of sound. Chapter 4 is comprised of a series of empirical studies in Austin, Texas, and Frankfurt am Main, Germany, that examine the effects of sound on the urban dweller and show how it influences the development of a city's sense of place in everyday life. This chapter includes examinations of noise perception and the nature of audio-guided tours, a study of sound-identity and the city's image creation (branding), and a consideration of the complex relationship between sacred sound and the creation of a sense of place. Chapter 5 considers the relationship between sound as media and the urban environment. The acoustic creation of a city in audio drama – acoustic productions developed from radio drama – can be incorporated into our environmental understanding and thereby play a role in shaping our perception of a city's sense of place. Two popular audio drama series – i.e., audio plays – are analyzed to demonstrate the effects of sound on the urban dweller's perception of the city. Chapter 6 summarizes the described findings and formulates conclusions. The theoretical frame presented in Chapters 1 and 2, combined with the methodological groundwork of Chapter 3 and the analysis of empirical findings in Chapters 4 and 5, demonstrates that any discussion about the urban environment must consider both the individual and collective experiences of sound. The acoustic element, just like the visual, plays its part in creating the city's sense of place.

Chapter 1: Thoughts on Sound and the City

We start with a philosophical discussion of individual perception grounded in the work of Edmund Husserl (see 1960, 1983, 1997, 2001, 2010) (section 1.1.1). Perception is always dependent on the individual's own abilities, which are fundamentally different from anything else. Bernhard Waldenfels' *phenomenology of the alien* clarifies the topic (Waldenfels 1997, 1998, 1999a, 1999b). Biographic and cultural context are part of the individual *stream of consciousness* that determines how something is actually perceived (see Husserl 1973). Ernst von Glasersfeld's *Radical Constructivism* (1995) adds the idea that if perception is understood as being highly individualistic, it becomes not a passive act but an active action (1.1.2). This point must be made clear from the beginning to avoid the notion that sound (in later chapters) is thought of as inherent to a city's sense of place *an sich* (in itself), as Immanuel Kant would put it (2007). It is perception that matters first and individual perception that creates the most problems when talking about sound in the city. Focusing on what is being perceived (1.2.1) we stay in the phenomenological tradition, as in the case of Yi-Fu Tuan's description of *sense of place* (1974, 1977). In addition to his mainly positive connotations, negative boundaries of place are also discussed (Relph 1976) (1.2.2), as well as parallel existing *sense of place* concepts that are individually driven and related to one and the same location (Cresswell 1996, Rose 1995). Rose's definition of the sense of place as "part of the systems of meanings through which we make sense of the world" (1995: 99) clarifies that position. Human geographical studies often use the *sense of place* concept (1.3.1). To get an overview of how urban geographers integrate that concept into their research, we take a look at various studies on the topic (Gregory 1995, Kianicka et al. 2006, Massey 1991). Theoretical thoughts and critiques on the concept (Shamai and Ilatov 2005) are also considered here, as well as a call for discussion about sound as part of a city's sense of place. In this context, sound as an element that is important for a sense of place is rarely considered. Existing research focuses more on the visual aspects of place. When sound is discussed, music and language are the center of the studies (Boland 2010, Leyshon, Matless, and Revill 1995, Pesses 2009) (1.3.2). While human geographical studies mostly spare sound when talking about individual perception of places, other humanistic disciplines use sound more often as an object of research (1.4). From art (Fontana 1987), ethnology (Sakakeeny 2010), and history (Gunderlach 2007) to musicology (Kun 2000), psychology (Boltz 2010), and sociology (Fortuna 2001), findings are presented that specifically focus on (urban) sound. As *sound* pushes to the center of attention the book finds its motivation to take a closer look at sound and the effects it has on the urban dweller.

Chapter 2: Sound Effects

Starting with human physiognomy we consider sound perception from an acoustic point of view and take a look into the process of hearing and acoustic perception (Byrne, Michael, and Tufts 2011, Clark and Cox 2012, Irwin et al. 2011) (2.1.1), continuing with a literature-based overview of the various effects sound can have on the individual (Gaver 1993, Morley and Somdahl-Sands 2011, Thurston 2013) (2.1.2). As the focus of this book lies predominately on the social and humanistic aspects of urban sound, its classification as *unwanted* (2.2) and *wanted* (2.3) are investigated. Starting with unwanted sounds, we look at that word's mildest form possible – annoying sounds (Kryter 1972, Lavandier et al. 2011) (2.2.1) – and then turn to its radical form, which distresses the human sense of hearing (Holmes 2012, Johnson and Cloonan 2009, Neustadt 2004) (2.2.2). In contrast to these wanted sounds, we learn about how a sound has to be shaped to have a positive effect on the individual. Frequency, duration, interval, loudness, and other elements will be taken into account. The findings are literature-based and vary from nice sounds like shopping music (Caldwell and Hibbert 2002, Fujikawa and Kobayashi 2010, Vanel 2008) (2.3.1) to acoustic ecstasy at live music events (Brazeal 2003, Duffy 2000, Jankowsky 2007) (2.3.2). The absence of sound also has an effect on the individual. Silence and sound create an interesting pairing in which each relies on the other (Bruneau 1973, Kania 2010, Mylott and Dubois Williams 2007). That the absence of sound is inherent in every perceived sound is discussed with the help of Edmund Husserl and especially Bernhard Waldenfels' *phenomenology of the alien* (2.4): "A phenomenon like the alien, which shows itself only by eluding us, could be characterized as a hyperphenomenon. Accordingly, Husserl characterizes the alien as 'a verifiable accessibility of what is inaccessible originally'" (Waldenfels 2011: 35). When we talk about the effects that sound can have on the individual, we refute the idea that these sounds are perceived passively and that the individual does not play her/his part when it comes to creating her/his own soundscape (von Glasersfeld 1989). Earphones are often used to listen to a personal soundtrack, which can have two aims: adding an individual acoustic touch to visual perception (DeNora and Belcher 2000, Tagg 1994) (2.5.1) or tuning out or negating acoustic surroundings (Du Gay 1997, Hosokawa 1984) (2.5.2). The findings of Michael Bull (Bull 2000, 2001, 2007) are of importance here, as are those of other studies relating to the impact of earphones on the environmental experience (Simun 2009). With the description of how *artificial* sounds are used by the individual to perceive the environment, the focus once again lies on individual active action. What section 1.2 discusses with a radical constructivist approach theoretically is thus accompanied by day-to-day examples. By describing possible effects we get a better understanding of urban sounds from different scientific perspectives. Before empirical research can begin, we have to formulate a methodology that allows us to gather information about sound, help describe the resulting effects, and formulate subsequent research questions.

Chapter 3: Sound in the City

The first empirical approach to investigating the meaning of environmental sound for humans was conducted by the World Soundscape Project, a Canadian group associated with Murray Schafer, in the early 1970s (Schafer 1994) (3.1). The soundscape – deriving from *landscape* and defined as "an acoustic environment" (Schafer 1994: 6) – was recorded in *audiowalks* or *soundwalks* (see Westerkamp 1988). "A soundwalk is any excursion whose main purpose is listening to the environment" (Westerkamp 1974: 1).

The soundscape can be deconstructed into single *sound events* (3.1.1). Taken out of their environmental context sound events become *sound objects* that are used for composition (Truax 2002). Schafer developed three categories to specify *sound events* (3.1.2). The *keynote* builds the base for any soundscape and includes vague sounds like traffic and wind, while a *signal* like the barking of a dog or the siren of a police car, can easily be identified. A *soundmark* is described as a community sound that represents a certain location or neighborhood, like the way the bells of Big Ben are associated with London, England. Schafer provides us with an evaluation method for translating all single sound events into words, numbers, and signs (Schafer 1994, Truax 1978) (3.2.1). This chapter shows how Schafer's evaluation method needs to be transformed and extended to meet the expectations of twenty-first century research methods. Topics like upgrading paper to digital data and mapping sound in cartography are addressed (Arana et al. 2011, Kornfeld, Schiewe, and Dykes 2011, Papadimitriou et al. 2009) (3.2.2). Depending on research focus and technological possibilities classification of single sound events varies (3.3). The meaning for and impact on the urban dweller can only be evaluated after establishing a viable classification system. A very individual approach comes from acoustic ecology, where classification is based on personal feelings and everyday knowledge (3.3.1). There is also the possibility to divide sound events into categories like *human, nature,* and *technology* sounds (see Schafer 1994). The variety of classification possibilities is shown via a literature-based comparison of urban planning with social science definitions (Lebiedowska 2005, Papadimitriou et al. 2009, Rychtáriková and Vermeir 2013) (3.3.2).

To utilize acoustic ecology for the *geographies of urban sound,* its approaches have to be customized. With the help of three empirical studies, such adjustments are discussed (3.4). In a field experiment I conducted in the city of Lisbon, Portugal, several soundwalks were recorded that show how single sound events can be deconstructed into keynote sounds, signals, and soundmarks. Finding examples for the latter is only possible by consulting the local population, who can tell what sounds qualify as soundmarks. I describe this with quantitative and qualitative data acquisition involving Lisbon's inhabitants (3.4.1). Another approach to raw sound recording is to record and categorize single sound events directly. A second empirical study conducted in London, England, demonstrates the possibilities of that approach for geographical empirical research (3.4.2). The third research project combines the first two and turns the focus to space with *sound point recording,* a new location-specific sound analysis I developed in Austin, Texas (3.4.3).

Chapter 4: Uses of Sound in the City

Urban sound appears in various situations. After a general overview (4.1) we start by looking at everyday approaches to *noise* (4.2). In city planning noise is simply classified as *unwanted sound*, but there are acoustic explanations that have to be considered as well, as interviews with experts indicate (Murray 2011) (4.2.1). Examples of noise prevention and mapping in multiple (European) cities are given, including findings from my own field studies. The impact of noise on the urban dweller is discussed with empirical data from interviews in Austin, Texas (Leasure 2011), backed by a collection of existing noise reports and official regulations (European Commission Working Group Assessment of Exposure to Noise (WG-AEN) 2006, Texas Department of Transportation 2011, US Department of Transportation 1995) (4.2.2). The radical constructivist position that defines noise in relation to individual perception (von Glasersfeld 1995) and the thesis that noise is based on personal biography (Husserl 1983) question regulations that understand noise as sound that "disturbs a reasonable person of normal sensibilities" (The City of Austin, Texas 1992: § 9–2–1 (4)(b)). As the findings from a large field study in German Rhine-Main Area about air traffic noise – including data from a standardized questionnaire and a large number of qualitative interviews – show, noise perception is strongly co-defined by individual experience and subjective perception (Balzer 2012, Heiligenthal 2012, Szafera 2012).

Sound can also have positive connotations and be used in many good ways, as in the case of the city of Austin, which uses sound to create a unique image in order to brand itself as the Live Music Capital of the World (4.3.1). Official statements from Austin's City Council, Convention and Visitors Bureau, and Creative Media Center are discussed (Reyes 2011, Vallejo 2011) and set into perspective when we analyze the local media, using the example of the city's daily newspaper, the Austin American-Statesman (4.3.2). The tourism industry has often used audio to attract more visitors (Tallon) (4.4.1). Section 4.4.2 presents different approaches to using audio guides in the city, using the example of museums in Austin. Findings from research in major museums are presented to show the different possibilities of audio-guided tours in museums (Cid 2011, Denney 2011, West 2011). In addition to this use of audio tours, the city itself can be explored with auditive guidance. Examples from different tours in Austin, Texas, show the potential and problems of such material (Lady Bird Johnson Wildflower Center 2009, Save our Springs Alliance 2010, Texas State Travel Agency 2012) (4.4.3), especially when a tour's stops are only connected by the listener's imagination. Individual associations that occur while the urban soundscape is actively tuned out (Bull 2000) support the radical constructivist approach.

We look at individual perception in the last section of this chapter, focusing on the abstract motive of the *sacred* (4.5). Interviews conducted in churches in the downtown area of Austin show the significance of sound to the sacredness of a place, where silence is of importance and where unique sounds can be found that in the urban dweller's perception represent sacredness (Heenan 2011, McDonell 2011, Stevens 2011). Husserl's "verifiable accessibility of what is inaccessible originally" (Waldenfels 2011: 35) helps explain why "sounds of the services" are used to "create an environment for silence" (Barr 2012).

Chapter 5: Excursus: The Medial Creation of the City in Audio Drama

This second empirical chapter takes us away from the sounds occurring in the environment and focuses more on the impact of sound, in this case the importance of images produced by the media for our perception of the world. While most research focuses on the visual impact of media, sound has been neglected by most studies (Adams 2009, Aitken and Zonn 1994, Rose 2012) (5.1). In comparison to examples given in sections like 4.5, sound is *brought* to the city via the individual's *stream of consciousness* (Husserl 1973). Sounds that the individual perceives while watching a movie or listening to an audiobook can have a direct impact on her/his perception of the city's sense of place when connotations and emotions are transferred from media to *actual* environment. To go more into detail and give examples of both the construction of the city in media as well as the potential impact on individual perception of the actual environment, I present my findings from a project about popular cultural audio drama genre, especially *The Three Investigators* series (Wissmann and Zimmermann 2010) (5.2.1). Unlike typical audiobooks, where a narrator talks to the listener, that series and mystery/thriller audio play *Gabriel Burns* use additional sounds, music, and specific voices for each character to tell their story. Having evolved from the radio drama, audio plays today are published on audio media like CDs or as digital downloads accessed via the Internet. As the world's most successful audio play, Germany's *The Three Investigators* draws a picture of coastal California that is overloaded with cultural stereotypes and generalizations. Gabriel Burns' cataclysmic description of Vancouver, Canada, explicitly does not focus on generalization but has a similar impact on its listeners. The psychological effects of certain sounds, as described in Chapter 2, and acoustic images help to direct the story and listeners' feelings. The audience can only follow the story if auditive decoding works, which means listeners have to be familiar with the sounds they hear. Otherwise, the audio content is "excluded from [an own] specific order" (Waldenfels 2011: 4). Necessary generalization leads to cultural stereotyping and a specific picture of the described world (especially cities) that influence the listener's expectation and perception of the *actual* (urban) environment.

A sound- and content-based media analysis of 70 audio play episodes shows how sense of place is evoked in audio drama (5.2). The three-part analysis starts by focusing on speech and associations with the city as place (5.2.1). The storyline is then divided into urban places of action (5.2.2) that are evoked not only through the audio play's dialogue and narrator (i.e., storyteller) but additional sounds (i.e., *keynote, signal,* and *soundmark*) (5.2.3). This helps answer whether or not there is a common method to acoustically create the urban environment in audio drama.

Chapter 6: Sound Is the City

This final chapter summarizes the findings of the book by integrating sound into the city's sense of place (6.1). The value of these new views of urban sound for the humanities is shown (6.2), with special focus placed on the output and benefits for human geography (6.3). Finally, sound is not the only missing sense in understanding a city's sense of place. A consideration of its integration into the urban fabric helps us understand these missing pieces as potential fields of research (6.4) that might open up even more options for urban studies than the focus on visual elements alone allows. This reflection forms the conclusion of Chapter 6.

Geographies of Urban Sound should help us realize that there is more to the urban environment than just the visual. If we want to understand processes and actions going on in the city, sound supplies us with input we must not tune out. So, listen up!

Chapter 1
Thoughts on Sound and the City

1.1 Individual Perception as Basic Requirement

If we think about sound and the different qualities that it can have, we have to realize that individual perception plays a major role in what we perceive and how we perceive it. While it seems to be quite obvious that what an individual hears, sees, and smells is due to her/his own senses, this issue has to be addressed before we can start thinking about sound as it affects us as part of the environment. Edmund Husserl, one of the fathers of phenomenology, studied the topic (Husserl 1992a, 1992b, 1992c) long before the sense of space and place became an issue in human geography (Relph 1976, Tuan 1974, 1977). Bernhard Waldenfels' work on the phenomenology of the alien is also of importance to the question of perception (Waldenfels 1997, 1998, 1999a, 1999b). In the process of perceiving and recognizing things the individual is confronted with other egos and alien elements that have to be assigned to the ego. Otherwise perception would never be an active act but rather a passive absorption.

1.1.1 Stream of Consciousness

"Incessantly the world of physical things and, in it, our body, are perceptually there" (Husserl 1983: 83). Phenomenology has always been about observing physical things. Edmund Husserl (1859–1938) started the philosophical sub-discipline with the idea that we should stop interpreting the world but instead use perception as the basic requirement to make statements about it. It is the physical world that can be perceived, so we should focus on things we can sense through individual perception. Perceptionally, the physical world is always there, as are own bodies, all other bodies, and things (Husserl 1999: 54).

> The world is the sum-total of objects of possible experience and experiential cognition, of objects that, on the basis of actual experiences, are cognizable in correct theoretical thinking (Husserl 1983: 11).

The physical world becomes the center of attention, but it is not the primary focus of Husserl's writings. The process of perception is far more important. There may be physical things that we perceive, but by looking at them, or, in our case, hearing them, all things become an "object of consciousness" (Husserl 1983: 310). They are simultaneously part of our consciousness and physical things. Or to put it another way: The world exists because we perceive it.

Perception itself is divided into internal and external perception. Internal perception concerns our thoughts, feelings, and "soul", while external perception is "our perception of bodies" (Husserl 2001: 336). This thinking goes back to British empiricism and John Locke, who philosophized about human identity based on physical perception. Of course, British empiricism is still driven by the quest to find evidence for the existence of god, but while Locke is searching for that evidence, he presents some findings that are relevant to the cause of external perception (Wissmann 2011: 31). While Plato, in his theory of ideas (Plato 2008), takes the "idea" as something metaphysical, Locke ascribes it to individual thinking:

> Every man being conscious to himself that he thinks, and that which his mind is applied about whilst thinking 'being' the 'ideas' that are there, it is past doubt that men have in there minds several 'ideas' […] (Locke 2000: 45).

Physical things have an impact on individual thinking and generate ideas that are linked to them (Husserl 2001: 159). But ideas are not passively perceived; they arise inside the human mind and actively bring sense to the world – a concept that will be discussed in the section of this paper dealing with the basis of radical constructivism (1.1.2).

Husserl calls this perception of external things "originary experience" (1983: 6). The only other experiences originary to us are our own body and anything internal, things recognized through self-perception. The common element between external and internal perception is sensuality (see Locke 2000: 78). Though we know about the meaning of perception, we haven't precisely named how perception works in a phenomenological way. It is rather self-explanatory:

> By my seeing, touching, hearing, and so forth, and in the different modes of sensuous perception, corporeal physical things with some spatial distribution or other are simply there for me, 'on hand' in the literal or the figurative sense […] (Husserl 1983: 56).

Husserl argues that external perception – keeping in mind that it is sensuous – is not deceptive but evident (Husserl 2001: 336). It is perception that enables us to experience the world. Without perception there would be no experience possible (Husserl 1983: 11). If we were not able to experience the world through perception, than perception itself – as a concept – would be useless. Whether perceived physical things point to ideas in Plato's sense or are (just) ideas of our mind, as Locke would argue, perception always equals apperception. "Outer perception is apperception […]" (Husserl 2001: 341) – which speaks to George Berkeley's *esse est percipi* (Berkeley 2009).

So, let us talk about continuing perception and how experience evolves. Husserl argues that if consciousness is given, it contains things that we recognize from former perception. Radical constructivists (see 1.1.2) would call this "object permanency" (see von Glasersfeld 1995), which describes any "recollections of consciousness [...] which lead to a stream of consciousness" (Husserl 1997: 167). The stream of consciousness contains identical elements or objects that are being perceived and each correlate with their former appearances. Thus, objects are not only recognized in their individual (physical) form but also as bearing all possible variations and appearances in different contexts or settings (Husserl 1992a: 186). And this is exactly what distinguishes Husserl's process of perception from Locke's. For Husserl, objects are not only immanent in the stream of consciousness as they appear while being perceived. Alternative versions and possibilities of being, active behavior, and passive existence – every possible state is incorporated in the stream of consciousness (Husserl 1992a: 123).

The stream of consciousness we are talking about is always an individual stream. This does not mean that objects have to appear differently inside each particular stream. Based on the finding that objects are linked to former appearances, each stream of consciousness is building upon a totally different setting of objects. Thus, there may be analogies between streams – and there are, if we think about common sense and public opinion – but each stream is individual:

> If we take, for example, the level of the perceptual physical thing-constitution, the correlate of which is the thing pertaining to the senses furnished with sensuous qualities, then we refer to a single stream of consciousness, to the possible perceptions of a single perceiving Ego-subject (Husserl 1983: 363).

Objects and everything else in the physical world are thereby existent because of being perceived individually (Berkeley 2009: 11). So, whose stream of consciousness are we referring to? As stated earlier, perception is not only external but also internal. That is why perception is linked to physical objects and, at the same time, to the act of perceiving itself. The *I* also perceives itself as being the one perceiving. "'I' perceive, specifically this or that" (Husserl 2010: 8). Here we are confronted with the presence of the *own*. Husserl scholar Bernhard Waldenfels later contrasts this presence with the *alien*. But before we discuss this pair we have to keep in mind that the *I* Husserl (as well as Waldenfels, later on) is referring to is thought of as being bodily. It is a "presence >in person<" ("leibhaftige Selbstgegenwart") *I* is conscious of (Husserl 1983: 83, 1992b: 81). Physical objects are the ones we can see, but other senses, like hearing, are also understood to be bodily:

> I hear something, namely the tone of a violin, the shouting of the children in the street, the droning of a bee. This everywhere. I also see and hear myself and other people; I see my hands, and I hear words and noises which belong to me, to my Body. Seeing and hearing relate in the first instance, even in the case of the perception of others, to what is Bodily (Husserl 2010: 8).

Now that we know that the body of the *own*, as one of the things that can be perceived, is part of the stream of consciousness (Husserl 1960, 1995: 102), its position in the world is of importance. Every object that is part of the stream is linked to the *I* or *Ego*. That does not mean that this center has to be the center of attention or importance, but rather that it is significant in a correlative way: *Here* is the ego and *there* is everything else (Husserl 2010: 109). What applies to physical things is also right in a psychophysical sense. There are other egos, too, that are part of the ego's stream of consciousness. These egos are thought of as having their own consciousness and perception but, as part of the ego's perception, they are sensed – non-originary – in relation to the centered ego (Husserl 1983: 124).

Husserl's writings (1960, 1990), and especially the works of Bernard Waldenfels (2000, 2004, 2005), illustrate that phenomenology has paid a lot of attention to the relationship between the "own" and the "alien" – and not only in a physical sense. The most important examination of this relationship takes place in Waldenfels's tetralogy on the phenomenology of the alien (Waldenfels 1997, 1998, 1999a, 1999b). Own and alien are concepts referring to one another. Alien can only be something in relation to something else, which is the own. And the own is the familiar, the usual, the known in our lives (Waldenfels 1997: 74–7).

> If we take the alien, in contrast, as something that cannot be pinned down, if we take it as something which seeks us out in our own home (German 'heimsuchen') by disturbing, enticing, or terrifying us, by surpassing our expectations and eluding our grasp, then this means that the experience of 'the' alien always affects our own experience and thus turns into a 'becoming-alien of experience' (Waldenfels 2011: 3).

Perception has an affect on our lives. All things we sense through apperception impact our stream of consciousness. Depending on how alien the encounter is, the more or fewer changes it brings to the stream. But the alien does not have to be something totally vague and nebulous. There are several degrees of alienness (Waldenfels 1997), though one thing is always true: the alien can never arise out of the own, nor be affiliated with everything general (Waldenfels 2006: 57). It is a quality of the alien that it is originary inaccessible (Husserl 1960). Thus, the bodies of other humans can be perceived originary (originally), while the self-perception of the alter ego stays alien: "The Bodies which are externally standing over against me are experienced by me in primal presence just like other things, whereas the interiority of the psychic is experienced in appresence" (Husserl 1990: 172).

While the alien is originary inaccessible, the own can be directly perceived (Waldenfels 1998: 42). It is ordinary in a sense that everything belonging to the own follows a certain (own) order. One's mother tongue is a good example of this. When learning their first language, children operate in a certain cultural context that is/becomes familiar to them. At the same time, all other languages that are not the mother tongue become alien languages, as they are not ordinary but extraordinary (Waldenfels 1999a: 184). Own is everything on this side of a border; the alien lies beyond. It is never possible to cross the border from this side or that. If the alien crossed it, it would immediately become extinct (Waldenfels 2002).

We have to be careful here not to think of this border too figuratively. It is no border between states or countries, no frontier between good and evil. Good and evil – in this sense – would be part of the own as something ordinary or known (Waldenfels 2006: 59). The alien can also be understood as the question that causes the own answer (Waldenfels 1998: 95–6). We never answer a question that we ask ourselves or that we could answer easily. There is always something alien about it, something unknown that we have to answer. The act of answering itself shows that the own and alien are deeply interlaced and cannot be thought of without each other (Waldenfels 1994). In the end, the spatial analogy that best correlates to own and alien would be the one dealing with *here* and *elsewhere* (das Anderswo) (Waldenfels 1997: 26). Which brings us back to the nature of *here*. We can only explain *here* with its *bearer*, i.e., the own whose *here* we are talking about. *Home* is a fitting opposite to everything alien, which is not home but someplace else. And again we have to be careful not to relate *home* and *elsewhere* to only physical space but rather to understand them as psychophysical concepts.

Waldenfels gives several examples of this when he describes the different degrees of increase to alienness (Waldenfels 2006). First, there is the *alienness of the everyday*: things that are incorporated into the own's stream of consciousness – indeed, there can also be alien elements inside the own. Take the mailman, for example. His body is something we can perceive originary, but his alter ego cannot be sensed. It is only recognized via apperception. Because the own consists of his body and ego, there has to be an alter ego for each other (human) body. Listening to a piece of music we have never heard before is another example. Sure, it is possible to immediately categorize a song as jazz, rock, or classical because we know how jazz, rock, and classical are supposed to sound (Waldenfels 1999a: 185).

Second, *structural alienness* is further away from the own than the alienness of the everyday. A foreign language is a perfect example of this. We are aware that foreign languages exist, even those we do not speak or understand (Waldenfels 1997: 36). A former zeitgeist may be part of the history of the country we live in. We know about it without having lived it (Waldenfels 1999b: 81). There is more than one cultural context on our planet. The own's society, with its rituals and traditions, is only one, and surely there are more we do not know about. A structural alien piece of music would be a piece that we could name as music, though it would be hard for us to name the style, and certain tones would sound disharmonious to us.

The third category, *radical alienness*, is the most extreme form Waldenfels discusses (Waldenfels 1999b: 115–16). In contrast to Simmel's *total alienness* – for example, inhabitants of Sirius (Simmel 1908: 509) or some sound we would never associate with music – radical alienness describes extremes like death and Eros: situations we know exist but cannot understand or empathize with without experiencing them. Music contains radical alienness, too. Sometimes, there is something more to a song than just its sound. It may be hard to describe what exactly this is that makes music affect us in a certain way, but still we are affected (Waldenfels 1999a: 185).

In one of these ways, alienness affects us, disturbs us, makes us stop. Whatever degree of alienness we are confronted with, it always refers to us as own. It is the question we have to answer, and it is the own that gets asked. In this respect, the alien always circles around a center of the familiar, of the known, of home. In the middle of it we find ourselves as ego (or I), which is in charge of how we handle all external (alien) impact and build up stream of consciousness. The one who is talking, seeing, and sensing is living in the center of its intimacy and familiarity. This center – like the border – cannot be thought of as being static in physical spatial terms. It is permanently surrounding the own and thus is always on the move without ever removing itself from its physical body (Husserl 1991: 281, 2010).

And there is a second thing we can learn from the phenomenology of the alien: The own perception – bound to the ego – first of all recognizes things common and non-alien.

> As Ego in the transcendental attitude I attempt first of all to delimit, within my horizon of transcendental experience, 'what is peculiarly my own' (Husserl 1960: 97).

The (own) ego perceives the world beginning with the familiar and known. That not only means that perception starts with everything own and then works its way through the different degrees of alienness. It also means that there is an own base to each object or element perceived, regardless of whether it's alien or not. Without having an own core or center of awareness and cognition, nothing (alien) could ever be perceived or even recognized (Husserl 1973: 432). The ego divides everything it perceives into own or alien, but – as shown above – the constitution of the own as well as the alien is causally determined by the own.

1.1.2 Active Construction of the World

As the phenomenology of the alien has shown, it is always the ego that is center to everything it perceives. The question remains whether external input is passive or if the ego does even more to all things perceived than centering them around itself – that those things could be actively constructed by the ego.

If we take a look at neurobiology we'll come across a belief in the active construction of the world. Humberto Maturana and Francisco Varela believe that experience is always linked to the observer, an assumption fitting into phenomenological thought concerning the ego and its stream of consciousness (Maturana and Varela 1992). The observer himself divides the content he perceives into single things and elements. Thus, the act of differentiation comes before things and elements are perceived as being a body, object, or other thing. What is:

> observed are not things, properties, or relations of a world that exists «as such», but rather the results of distinctions made by the observer himself or herself.

> Consequently, these results have no existence whatever without someone›s activity of distinguishing (von Glasersfeld 1997).

Jean Piaget (1992, 1997, 2007) did a lot of research on the perception and consciousness of children. In psychology and pedagogy he is often cited for his work in developmental psychology. The young mind is supposed to show best how cognitive processes work and how the individual constructs his/her world. Cognition never equals reality, and even the originary perception of external things is always bound to existing cognitive structures (Piaget 1974). Cognition has to be thought of as an assimilation of new contents of meaning into former elements of the own stream of consciousness. "Piaget [...] posits as primary reception the active mind that organizes sensation to form percepts" (von Glasersfeld 1995: 111). Experience, in this context, derives from adjusting the things perceived to known ideas of them – a process Piaget describes in four phases.

The first phase is *assimilation.* According to Husserl and Waldenfels, whenever there is originary perception of external objects, bodies, or things, these elements are integrated into the stream of consciousness (Husserl 1992a, 1997, Waldenfels 1997, 2006, 2011). Piaget uses the term *assimilation* to illustrate the integration of something external into an existing structure.

The second phase is *perturbation* (Piaget 1976: 131). Of course, assimilation only works if perception matches expectation. If the individual perceives something that does not quite fit to his existing image of it, a perturbation occurs. Take an apple for example. That object is defined through former experience as eatable, roundish, tasty, and red. The appearance of a green apple provokes perturbation. The image of an apple being only red doesn't fit with the perception of a green one.

The third phase is *accommodation.* To solve a perturbation, an act of learning has to take place. The existing image of an apple has to be modified to fit the perceived external thing *apple* with one's own stream of consciousness. The phenomenological term *stream* is appropriate here, as it shows the changes possible to objects inside our consciousness over time.

The fourth phase is *equilibration.* After accommodation is done perturbation is solved (Piaget 1976: 11). From that point on, the *apple* will be known as something eatable, roundish, tasty, and red *or* green. There is always the chance that perturbations do not get solved and that the deconstruction generates even more perturbations, maybe in another section of the stream of consciousness.

Although all perturbations might be equilibrated at some point, perception of the real world is never guaranteed. Heinz von Foerster (2000) explains this with the example of the blind spot in the human eye. If you close one eye and keep the second one open, you are unable to see an object like a bullet on a page from a certain direction and angle. This shows that there might always be something our senses cannot spot, something bearing a potential perturbation for our experience and construction of the world. Like the phenomenological ego, when we are talking about perception we always mean the perception of one individual being.

Maturana believes that insofar as each human is a singular entity, perception is always different (Maturana 2001: 17).

Thus, the world appears differently to each individual, depending on the individual stream of consciousness, definitions of everything perceptible, and attributes of the same – depending on perturbations and equilibrations.

The differences between people's perception become obvious in the intercultural context. Psychologist Paul Watzlawick describes what he calls the *nonsense of sense or the sense of nonsense* (Watzlawick 1995). Cultural settings often determine how we perceive, process, and act on external information. There is a wonderful example of a riding club in the Brazilian city of São Paulo. An American is talking to a Brazilian and they are both trying to find the right physical distance to each other while talking. In the US people usually stand an arm's length from each other during an interaction. In Brazil people stand much closer. So, during the conversation the Brazilian keeps stepping toward the American, while the American keeps taking the same step back to re-establish the right distance. This goes on for a while until the American falls from the porch and hurts himself (Watzlawick 1995: 23–4).

Though this story is meant to be funny, and is built on strong stereotypes, it also shows how day-to-day situations are determined by our own perception and the (mis)matching of our behavior with others. Reality, according to Watzlawick (1998: 7), is the result of communication. What is real no longer belongs to the sphere of Platonic ideas but the perception and interpretation of the individual. It is not the truth that matters for the individual construct of the world to work, but *viability*: "To the constructivist, concepts, models, theories, and so on are viable if they prove adequate in the contexts in which they were created" (von Glasersfeld 1995: 7). To achieve viability all perturbations have to be equilibrated. As reality as such does not matter, each working concept of an individual reality differs because of the unique ways perturbations are solved. The stream of consciousness allows the individual/ego only to construct/perceive its world in a certain way.

In the language of radical constructivism, the stream of consciousness is explained via *object permanence* (Piaget 1974) and *proto-time* and *proto-space* (von Glasersfeld 1995). If we see a ball disappear behind a wall, and see it again on the other side of the wall, we don't doubt that this is the same ball. Instead, we were just not able to see it for a while because of the wall. But Piaget (Piaget 1976) points out that babies, until they reach a certain age, are not able to follow the rolling of the ball. After they see it disappear their eyes do not follow the imaginary path behind the wall. The fact that the assumed rolling of the ball while out of sight is indeed imaginary – because we do not have any evidence that there is actually a ball rolling – brings Glasersfeld to the conclusion that we construct time and space actively to make it possible for the same ball to reappear (von Glasersfeld 1995). Otherwise we would have no evidence that it is the same ball. Proto-time and proto-space illustrate how the human mind actively constructs the world. It is our invention (von Foerster 2000). As Ernst von Glasersfeld writes:

1. a) Knowledge is not passively received either through the senses or by way of communication;
   b) knowledge is actively built up by the cognizing subject.
2. a) The function of cognition is adaptive, in the biological sense of the term, tending towards fit or viability;
   b) cognition serves the subject's organization of the experiential world, not the discovery of an objective ontological reality (von Glasersfeld 1989).

We do not have to follow Glasersfeld or classify the world a *total individual construction* to understand that perception is always founded in the individual or ego. Whether all objects perceived are the invention of a creative mind or exist independently, it is our eyes and our ears that sense the world. Equipped with our own personal stream of consciousness, where all things perceived are stored and updated, we expect objects and places to appear in a certain way.

1.2 Sense of Place: Concept and Multidisciplinary Theoretical Approaches

Places are where all things we perceive appear and can be sensed. The urban environment, as a surrounding where most people on the planet live, contains "many features and experiences", as Doreen Massey writes (2000: 6), adding that "simply cataloguing these has not proved enough to define the specific qualities of the city". In this formulation we find phenomenological, rather than constructivist, truth, as there seems to be a general thinking that there is an existing physical world where all appearances take place. Aside the ontological debate, Tuan (1974) recognizes that every individual perceives her/his environment in a different way. This is because the cultural background, socialization, education, gender, age, and socioeconomic situation of each individual varies. The environment itself isn't affected by this highly individual perception. It is not thought of as an active construction, even though perception is understood as being active (Tuan 1974: 12). The world is always the same, and it is only interpreted or understood in a unique way: "Taking the physical setting as uniform and constant, we saw how people with different experience, socioeconomic background, and aim evaluated it" (Tuan 1974: 74).

You could say that experiencing the urban environment and sensing its unique place is not only an individual or social phenomenon – first of all a city has to be built – but it is also embedded in the physical place. Surroundings have an effect on the urban dweller (Banerjee and Southworth 1990: 771). This does not mean that the effects derive from the physical piece of earth itself, but that place can be designed to have a certain effect on people. "The physical environment itself has an effect on perception" (Tuan 1974: 246).

On the one hand, individual/environment relations are based on place attachment, signifying emotional bonding to a place. On the other hand, there is place identification, which impacts the own physical body as being part of a place

(Rollero and De Piccoli 2010: 199). Before we examine both place attachment and place identity (see 1.2.2) let us stay with perception and individual experience for a while.

The emotional and physical impact of space on the individual shows that the urban dweller perceives space in several ways: individually, biologically, and socially (Tuan 1974: 245). "To be human is to live in a world that is filled with significant places" (Relph 1976: 1). These places are always perceived differently (Relph 1976: 57, Tuan 1974: 5, 12, 246, Tuan 2001: 43). As it is obvious that the social aspect of perception falls under the primacy of subjectivity, the question remains: How does the individual experience place? A very practical explanation is the one Tuan presents: "With hearing, and particularly with seeing, we seem to be actively exploring the world beyond us and getting to know it objectively" (1975: 152). It is quite interesting to see how individual perception has the potential to reveal the real world – in an ontological sense – but it first shows the problem with places as a whole. Especially when it comes to cities and other *points* you can locate on a map or in an atlas, individuality doesn't seem to matter. Latitude and longitude do not change. But the feelings toward a place open a whole different aspect of place experience. Thinking about your home town will trigger other feelings than thinking about just any city you have never been to before, or a city you've only experienced through media consumption, which actually is another important element of place experience (see below).

Furthermore, Tuan's idea of "exploring the world" divides the biological senses into more and less useful ones. When we talk about urban sound it is relieving to have hearing defined as being part of one's "active experience", but the number one sense clearly is sight. "Settlements are places because they are not only experienced passively but also can be seen [...]" (Tuan 1975: 159). From a constructivist point of view the split between seeing and passive experiences like smelling or touching is arguable. As Tuan himself concludes: "What can be known is a reality that is a construct of experience, a creation of feeling and thought" (Tuan 1977: 9)

Whether all senses should be understood as being of equal value for the experience of place or not, there is no doubt that the world is actively perceived via the senses. We are living in a world with places that can be understood as space with added value (Tuan 1977: 6, 136), a world that is of such high diversity, we are not capable of fully understanding it (Relph 1976: 6). But the phenomenological approach to describing what we sense is a valuable way of approaching the topic (Relph 1976: ii). The question remains what kind of value space can have, and what it might look like.

1.2.1 About Place

The city is place. Though we might take the city for granted as an objective entity, it is a product of individual perception: "People live in the same city, even in the same part of the city, and yet perceive different worlds" (Tuan 1974: 248). A city is a good example of a medium size place – as compared to small places like beds or huge ones like countries – that is expected to have special meaning. We can find it on maps, visit it, read about it in the news, and recognize it by its name (Tuan 1975: 157). Although cities are well known, there are several downsides that make them harder to separate from similar settlements. The lack of visible identity (Banerjee and Southworth 1990: 90) is one of them. It refers to the sense of seeing – think about famous skylines, buildings, or the shape of the terrain. If the visual image is hard to identify, it is hard(er) to experience the unique *sense of place* of a city. Illegibility of the urban environment is another downside. If the urban dweller cannot understand what makes a city special, if s/he fails to deconstruct its very sights, signs, and symbols, the city will never be experienced as a distinct place. The same goes for smaller entities like neighborhoods, which often "lack sharp, physically defined boundaries, and [...] have no distinctive skylines that can be seen from vantage points outside them" (Tuan 1975: 157).

"[P]lace is whatever stable object catches our attention", Tuan writes (1977: 161). This theoretical description suits various places, not only cities but also the neighbourhood, house, or bed, the county, country, or the entire planet (Tuan 1975: 153). Experience binds the individual who is perceiving place to the very same. With memories, associations, and connotations place gets expanded to infinity (Addyman 2010: 112) but at the same time stays rather static. Places always have individual meaning (Tuan 1977: 179), but meaning is not changing regularly: "Place is a pause in movement. [...] The pause makes it possible for a locality to become a center of felt value" (Tuan 1977: 138).

So what kind of value can place have? Entrikin (1991: 6) speaks of an "areal context of events, objects and actions", while Relph describes it as follows:

> In our everyday lives places are not experienced as independent, clearly defined entities that can be described simply in terms of their location or appearance. Rather they are sensed in a chiaroscuro of setting, landscape, ritual, routine, other people, personal experiences, care and concern for home, and in the context of other places (Relph 1976: 29).

There are several descriptions of place that refer to its social value (Cresswell 1996: 3, Halpenny 2010: 409, Kyle and Chick 2007: 215). A change in perception due to cultural background can be observed in tourism. Guided city tours, for example, are often adjusted to the visitors and their country of origin when certain stopping points are explained (Dahles 2002: 784, Kelner and Sanders 2009: 138, Macdonald 2006: 119). Places like "the boss' desk" or "the secretary's desk" only exist because of their social construction: "'place' combines the spatial with the social – it is 'social space'" (Cresswell 1996: 3).

Other authors focus more on the impact of the individual (Entrikin 1991, Mitchell Pierce and Short 1994, Relph 1976) as a fundament of social experience. Every reader of a children's book, for example – to add the impact of media to the discussion – will get a different image of the places described in each story.

Furthermore, descriptions of place always vary from one story to the next (Mitchell Pierce and Short 1994: 490).

Aside from individual and social perspectives, place is also defined through scale. First, there is location size – we already talked about places like the bed, the city, and the planet. But besides size, sense of place consists of a variety of meanings from both local and global scales (Rollero and De Piccoli 2010: 88). In addition to regional, national, and international elements, home – on a very small scale – has a huge impact on the construction of place (Kyle and Chick 2007).

Overall, place doesn't have one single, consistent meaning. Place differs. Whatever scale or context, meaning is never inherent to place. Rose summarizes that place – i.e., the sense of it – is understood as being either naturally existing, socially constructed (as through power relations), or individually (personally) made: "Many geographers, sociologists and anthropologists […] assume, however, that places have no inherent meaning, only the meanings humans give to them" (Rose 1995: 98).

Perception and experience do not inevitably deny a physical fundament. In a phenomenological sense location is even indispensable. Nevertheless, the force of active perception and social meaning calls attention to the sense of place. Who is responsible for its social creation? "The urban environment is constructed around a set of 'appropriate' places, areas imbued with sets of meanings deemed correct by dominant groups in society" (Cresswell 1996: 47). In this Marxist approach, social construction is understood as power relations inside a society. But if the meaning of place changes only slowly – we talked about the static element before – this does not mean that place couldn't have different meanings at the same time. Place can be reinterpreted and detached from the ideas of the majority. Cresswell (1996) uses the example of graffiti artists in New York City to illustrate how the meaning and use of public urban places can be changed. Skater culture and the Parkour movement have also modified urban places, turning handrails and park benches into barriers to master and walls and garbage containers into hurdles to jump.

Places, in their static and yet multi-dimensional way, are concentrated points in space and elements of order. They contain a concentration of "our intentions, our attitudes, purposes and experience" (Relph 1976: 43). There is more to a place like home than physical elements:

> Home is an intimate place. We think of the house as home and place, but enchanted images of the past are evoked not so much by the entire building, which can only be seen, as by its components and furnishings, which can be touched and smelled as well […] (Tuan 1977: 144).

Looking ahead to a more multi-sensory city's sense of place, sound has to be added to the discussion. Think about the church bells on the opposite side of the street you hear every Sunday while lying in bed. Think about the giant cuckoo clock in the dining room that divides up silence all over the house with its monotone, cold ticking. Think about live music in downtown bars that vitalizes a whole urban quarter with creative potential. Speech, as a subcategory of sound, plays its part when place is communicated. The act of naming a place gives meaning to it. Holy places like churches are transformed by speech from being common buildings to houses of God.

> Naming is power – the creative power to call something into being, to render the invisible visible, to impart a certain character to things (Tuan 1991: 689).

All of our senses help us construct places. Whatever tools – like speech – we use to implement place, it is already there in its physical form as something that can be evoked. Place is, in a way, necessary for our lives (Cresswell 2004: 33), as – to go back to phenomenology – everything we perceive is bound to our own body (Husserl 2010: 8). With the fact of physically living in place and perceiving it comes a dual approach to place. On the one hand we are able to observe place as an external entity, filled with meaning and physical being. In the case of the city, we can walk through it, watch it on TV, and read about it in a common travel guide. On the other hand, we are part of the place. We identify with it, are part of its being as urban dwellers, and we create it with our beliefs and images (see Entrikin 1991: 7). This dichotomy applies not only to places on the city scale but to smaller ones like the home and larger ones like a country as well (Relph 1976: 141). Entrikin speaks about "Betweenness of place", where the individual is part of a world it has not completely constructed on its own:

> In order to create room for [...] a subject [/individual] we require two irreducible parts to the concept of place: place as the relative location of objects in the world, and place as the meaningful context of human action (1991: 10).

Places function as centers of experiencing the world (Relph 1976: 141). Individuals and groups are bound to them in sympathy, even love (Tuan 1974: xii). In the first works on place and *sense of place* we do not read about negative connotations thereof. It is always the positive aspects that affect us. Tuan (1974, 1975, 1977) sees love as the link between place and the individual. If perception helps to shape place and defines how we, as bodily beings, are physically connected to place and are an inherent part of it, it would seem improper to have negative feelings for, strictly speaking, ourselves. The possibility to connect to a place via negative connotations is also known. At least since Relph wrote about "placelessness" in 1976:

> Indeed our relationships with places are just as necessary, varied, and sometimes perhaps just as unpleasant, as our relationships with other people (1976: 141).

This possibility will also be addressed in the following section with Edward Said's description of the Orient (Said 2003). Before we talk about this alien impact on place, we will continue with our description of its sense in all different forms.

1.2.2 A New Concept

To describe place many theories have been developed that try to cope with the different elements constituting it. Focus always lies on a certain aspect like individual perception or social construction, depending on the theoretical model. But whether concepts examine *place identity*, *place attachment*, *place affect*, or *place dependence*: they all care about the relationship between the individual and place. Combining most of these ideas in a very generalized way is a concept that goes back to the 1970s in geographic research: sense of place (Jorgensen and Stedman 2001: 233). The "sense of place – is the phrase used by many geographers when they want to emphasize that places are significant because they are the focus of personal feelings" (Rose 1995: 88).

In his often-cited book *Topophilia* (1974) Tuan's initial description of place is positive. In one example he presents a picture of a blank skull in the morning desert. The skull is not meant to suggest death or danger, however, but rather to add to a romantic scenery as experienced by the observer (Tuan 1974: xii). But generally speaking, sense of place "is the meaning attached to a spatial setting by a person or group" (Jorgensen and Stedman 2001: 233). Sense is ascribed to place "by people, both as individuals and in groups" (Rose 1995: 88), and it is based upon our beliefs and feelings, our individual and social behavior (Jorgensen and Stedman 2001). There are some arbitrary categories surrounding sense of place (see Mitchell Pierce and Short 1994: 491), and it can be applied to most special research contexts, like sense of place and folklore (see Ryden 1993: 59).

In some way sense of place is an everyday term that is easier to experience than conceptualize. Tuan describes it as being more in the *bones* than the mind of the urban dweller (Tuan 1975: 157). What sounds a little vague here makes sense if we adopt this view to sound. Most of the time we consume, i.e., hear, urban sound, we do not listen actively to it. It is part of the urban environment and, therefore, taken for granted. If we assume that the sense of hearing, as all other senses, plays an active part in perceiving, experiencing, and therefore creating place, we can begin to understand that sense of place indeed is more in the bones than the mind. Pam Harvey talks about the atmosphere of a hospital (2009) in a way that explains how individual perception matters for sense of place. Old stories about the hospital inspire Harvey's imagination of the place and add to the sense of it: "I found the place comforting, despite – or maybe because of – the reality of what its walls had seen" (Harvey 2009: 286). Place history stimulates sense of place creation, as does the socio-cultural context in which place is embedded (Rose 1995: 89).

In times of globalization, sense of place is created by more than the individual and local groups. Mass media has a large impact on places, especially if they are of importance for a larger group of people than the family. Though not necessarily true for places like home, from city scale onward, sense of place finds a global impact. Local reference is harder to be pointed out (Massey 1994: 146) if sense of place is created by multiple global sources. If globalization is not only to be found in the economy but also in sense of place, the significance of local boundaries may decline. Reference to sense of place gets extremely variable and sometimes even hard to identify. Rose sums up the descriptions and definitions in five points:

- Senses of place can be expressed through a diversity of media.
- Senses of place can be established at different geographical scales, and these can intersect.
- Senses of place relate to identity in different ways: they may invite identification with a place; they may establish identity by offering a contrast between the place that is 'home' and the place that is 'away'; or the meaning of a sense of place may be Irrelevant to how people identify themselves.
- Senses of place may become more intense when they are perceived as being under threat.
- The same location may be interpreted through different senses of place (Rose 1995: 97).

Sense can also be eradicated. Relph (1976: ii) finds one example for it in the mainstream architectural concept of the 1960s and 1970s. Modernity, with its cubic, standardized forms and its tendency to ignore or even destroy historical elements, brings a radical change to landscape. The reconfiguration of landscape is described as "placelessness" (see also Casey 2009: x). It is a certain sense, not the place itself, which is destroyed. Redefining a place – as modernity and postmodernity afterwards did – means eradication of a certain interpretation. The newly created sense of place might not be accessible to individuals or groups who were accountable for the former configuration. Sense of place thus reveals social distance between those individuals or groups who can understand its meaning and those who cannot (Cresswell 1996: 146). Rose (1995: 92) refers to Edward Said's work on orientalism, wherein the orient as alien place helps to define the western world: "[…] the Orient has helped to define Europe (or the West) as its contrasting image, idea, personality, experience" (2003: 1–2). In constructing the orient as the opposite of the west (Said 2003: 58) a European sense of place is created through negative classification, a practice Waldenfels used to describe the terms "home" and "elsewhere" (2006). Everyone who is not able to experience the common sense of place is an outsider, a stranger (Tuan 1986: 12). The outsider is out of place, must not be trusted, and does not play by the rules (Cresswell 1996: 154). There is a remarkable analogy in this statement to Waldenfels "disturbance of the alien": It is not fear like the one of an enemy but more of the unknown – an unpleasant feeling, that a place does not feel familiar. The presence of an alien sense of place is perceptible but can never be deconstructed or be fully understood (see Waldenfels 1997: 42). The alien sense is a totally different approach to deconstructing place. Thus, places become extraordinary and can even appear scary or threatening (Mitchell Pierce and Short 1994: 494): The strange "disturbs and disrupts; it resists our grasping ego; it frustrates our desire to place, classify, or make consoling forms" (Tuan 1986: 19).

Because of the incompatibility of different creations, sense of place stands for a certain worldview, way of thinking, and collection of beliefs (Cresswell 2004: 11). Places create difference: The ascribed meaning of a place is important to the people using it. There are certain standards concerning what is and is not allowed. There are rules, norms, and expectations for a place that must not be ignored. Thus, place is defined and differentiated from other spaces. Sense of place reveals who uses place in the *right* way. This social definition of place divides own from alien: Anomalies in behavior are identified as other or alien, not belonging to the place (Cresswell 1996: 154, Rose 1995: 105). In the end, every sense of place – be it own or alien – tells us about ourselves. And yet, it is not the alien sense of place we are out to find. The extension of familiarity (Tuan 1986: 16), the well know, and the ordinary is important in order for us – as bodily beings – to experience the betweeness of place: "We gain a sense both of being 'in a place' and 'at a location'" (Entrikin 1991: 134).

There are different aspects of the sense of place that cannot sufficiently be described by Entrikin's betweenness alone. The phenomenological part is covered by place experience, which states that place is a construct of experience (Relph 1976: i, Tuan 1975: 165, cf. Husserl 1983: 11). In contrast to this aspect – that we have discussed in the previous section – there is place attachment. This concept is less epistemologically driven than place experience and can be divided into three major categories: place affect, place dependence, and place identity. Place dependence deals with power relations (Cresswell 1996: 47), while place affect describes individual emotions for a place (Halpenny 2010: 409). Place Identity focuses on cognitive structures that foster social and individual identity related to place (Rollero and De Piccoli 2010: 198), a category of place attachment we will look at shortly. Place attachment itself is about various social and individual links to a place, like emotions (Halpenny 2010: 409). "Place attachment can be defined as the affective link that people establish with specific environments" (Rollero and De Piccoli 2010: 198). Thus, place attachment describes pro-environmental behavior (Halpenny 2010: 410, Kyle and Chick 2007: 209) and deals with "how individuals construct their own identity through their residential [...] histories" (Raymond, Brown, and Weber 2010: 433). Especially in the field of psychology, this opens up even more specified investigations like *nature bonding* and *social bonding* (Raymond, Brown, and Weber 2010: 422).

> Place attachment is a multifaceted and complex phenomenon that incorporates different aspects of people-place bonding and involves the interplay of affect and emotions, knowledge and beliefs, and behaviours and actions in reference to a place (Rollero and De Piccoli 2010: 198).

Social interaction is important for the individual to connect to a place. Demographics matter for place attachment, while length of residence, for example, is supposed to have no effect (see Rollero and De Piccoli 2010: 199–203). This seems rather unlikely, as social issues, which are always of importance, are interlinked with concepts like occupancy and age. In Chapter 4, section 4.2, I will discuss the matter using the example of air traffic noise.

While place attachment implies the presence of somebody who is attached to place, i.e., a social group or single individual, place identity might be understood as inherent to a place. This is not the case. There is no ontological or otherwise *real* identity of a place (Relph 1976: 62). It is always based on a group's or person's relationship to place. "Individual images" are modified due to socialization (Relph 1976: 57) to fit an intersubjective place image. Thus, physical being and social ascriptions both join individual experience. And it is physical elements in particular that help to shape place identity when we think about cities. Landmarks like the Eiffel Tower in Paris help to form a general picture of the city (see Tuan 1974: 247). Establishing shots in films use landmarks to set the scene and mood for a story (see Rose 2012), so place identity is created not only via uniqueness but also in contrast to other places that cannot contain the same distinguishing marks. But like sense of place, identity is not a static structure and can vary over time and from one social context to another (Relph 1976: 62). In contrast to the overall concept, place identity seems to be mostly taken positively. Rose describes it as "a feeling that you belong to that place. It's a place in which you feel comfortable, or at home, because part of how you define yourself is symbolized by certain qualities of that place" (1995: 89).

Whatever subcategories of sense of place are described, all of them have on thing in common: When it comes to physical boundaries that link us to a specific place, visual elements seem to have by far the most striking effect. "Of the traditional five senses man is more consciously dependent on sight to make his way in the world than on the other senses. He is predominantly a visual animal" (Tuan 1974: 6). Sight is ascribed special meaning when it comes to navigating the urban environment (see Cresswell 1996: 154). Seeing is accepted as true while the sense of hearing, for example, is considered "'hearsay' or 'rumor'" (Tuan 1974: 10).

It is remarkable that Tuan points out the predominance of seeing (1974: 245) while at the same time elevating sound to the status of most affecting sense (1974: 8). He points out that language is highly linked to the quality of a city (Tuan 1994: 144). So, we can ask whether sound might have the power to create places. Maybe it is more than just an add-on to sight and must not be underestimated when sense of place is discussed.

> Sound can add so much to the ambience of a place. The cry of sea gulls and the tinkling of tram bells are San Francisco (Tuan 1994: 150).

In this quote we see the fundamental problem with understanding sound in relation to place: It is described as both adding "to the ambience of a place" and being place itself – "the […] bells *are* San Francisco" [my emphasis].

To investigate sense of place and sound even further, the following section starts with the geographic view on the topic while the closing section will talk about sound in the humanities collectively.

1.3 Geographical Studies on Sense of Place and Sound

Human geography frequently uses sense of place in its various subfields. As section 1.3.1 will show, the concept can be applied to studies about mobility and tourism, be it an examination of travel and the unequal distribution of resources in a globalized world (Massey 1991), a description of differences between tourists and locals in a European holiday destination (Kianicka et al. 2006a), or the depiction of religious-driven travel on the *Camino de Santiago* (Graham and Murray 1997). Geographies of media make use of sense of place when imaginative geographies in writing are deconstructed (Gregory 1995, Hausladen 1995) and movies (Murphy 1996) and electronic media are explored (Meyrowitz 1985). Sometimes similar concepts incorporate sense of place, like Schnell's (Schnell 2000) work *The Kiowa Homeland in Oklahoma*; Kennedy, Seil, and Zube's theoretical approach to landscape aesthetics (Kennedy, Sell, and Zube 1988); the characterization of the relationship between Canadian First Nation peoples (Windsor and McVey 2005); and research concerning "similarities and differences between Israeli-born and immigrant groups" (Shamai and Ilatov 2005).

Some geographical studies on sense of place deal explicitly with sound as it can be perceived individually (Rodaway 1994). Sound can also be part of cultural representation (Jazeel 2005), transported via language (Boland 2010), or signify social constructs like *tranquil space* (Matless 2005). Johansson and Bell (2009a) mostly deal with music as part of sound. Music has quite a bit of meaning in the construction of place (Leyshon, Matless, and Revill 1995), for example the Los Angeles of the *Red Hot Chili Peppers* (Pesses 2009), or lyrics by *Tom Waits* (Pesses 2011). Listening to music is examined in cultural context (Curti and Craine 2011, Romig 2009) and referenced to local social and physical surroundings (Che 2009, Daynes 2009, Duffy 2000, Duffy and Waitt 2011, Kuhlke 2009). Sound shapes places for the tourism industry as certain genres or music styles evoke specific physical landscapes (Connell and Gibson 2009) or places of pilgrimage (Alderman 2009). Music genres themselves are linked to place. They can derive from place (Graves 2009, Johansson and Bell 2009b) or shape its people and environment (Finn and Lukinbeal 2009, Johnson 2011, Lindenbaum 2009). The potential of media for sound in relation to sense of place is discussed by authors like Kruse (2009), who highlights the effects of the Internet on independent music. Radio, as a connector of diasporic societies, is analyzed by Keough (2009, 2011). Among other features, music even can have a political influence on place (Jackiewicz and Craine 2009, Kruse II 2009), demonstrating its deep connection with place.

1.3.1 Sense of Place in Use

> Much of what is written about space, place and post modern times emphasises a new phase in what Marx once called 'the annihilation of space by time' (Massey 1991: 24).

"Time-space compression" is a term describing the ongoing shrinking of a globalized world in terms of communication and travel. Within these parameters it becomes difficult to sustain sense of place grounded in physical elements and location. The dynamics of a shrinking world oppose a concept that Tuan has described as rather static (1977: 179). Sense of place describes the possibility to preserve identities and keep them from vanishing in the rapid development of globalization. It is about finding some rest and confidence in the own identity with the help of place (Massey 1991: 26). For sense of place to endure in times of time-space compression, the concept needs to be defined by its social elements only (Massey 1991: 27).

Due to today's extensive and inexpensive traveling, the touristic potential of place has become more and more important. Using the example of the "Alpine village of Alvaneu in Central Grisons, southeastern Switzerland", Kianicka et al. (2006a: 56) illustrate how sense of place differs between groups. Elements like landscape, social relationships, culture, and the economy differentiate between locals' and tourists' sense of place. The study shows – not very surprisingly – that tourists recognize the value of tourism as an economic factor for the region and don't link it too much to sense of place. Traditional agriculture seems to be far more important. Locals are also aware of the touristic potential but at the same time are afraid of the change it might bring. Soft tourism doesn't sound overly effect the sense of place, while mass tourism does. Theoretically speaking, this is the difference between a static and a dynamic structure. And dynamics are said to be harmful to a strong sense of place that grounds itself in the history of the place (Kianicka et al. 2006a: 61, also see Massey 1991: 27). It is not surprising to see individual biography extremely to sense of place. As for the village in the Swiss Alps, locals bind their own history to the sense of the place, while tourists point out esthetic landscape values. Overall, the varying answers given within the groups and the references to the own biography prove that sense of place is created individually and always depends on social context. As a tourist, I normally do not picture my own cultural heritage while looking at beautiful scenery in my holiday destination (Kianicka et al. 2006b).

Sense of place does not only differ between certain social groups but also different times (see Harvey 2009). Graham and Murray analyze the "diversity of overlapping perceptions and meanings characteristic of past and present pilgrimage" using the example of pilgrims on the *Camino de Santiago*, running through the European Union (1997: 389). Against religious beliefs, pilgrimage is defined as a social construct. Pilgrimage is considered to be heterogeneous per se (Graham and Murray 1997: 390). Though pilgrims are on a religious path toward Santiago de Compostela, their reasons differ. Thus, their experiences will never be identical; after all the *Camino* has multiple routes, each with a different length. Citing the European Council, the authors show that the *Camino* today has an image of not only being a religious trail but also a European one. It connects different countries all over the continent and is a symbol of "European integration" (Graham and Murray 1997: 399). It also offers touristic value. Landscapes and historical artifacts become tourist places through media and tourist perception (see

Kianicka et al. 2006a: 62). Thus, sense of place vis a vis the *Camino* has become extremely heterogeneous, even though from the pilgrim's point of view its sense of place may not have changed.

Other human geographic subfields like media and imaginative geography also work with the sense of place concept. Sense of place in media is not only a product of social and cultural conventions but also of the individual perception of the producer. Gregory (1995: 30) illustrates how the physical presence of the writer matters in travel literature. Travel writings like those of Nightingale and Flaubert are based on what the authors saw and read. They are a mixture of description and media analysis of existing sources. No empirical data was gathered via direct conversation in the countries the writers traveled to (Gregory 1995: 51). This demonstrates the imaginative quality of travel writing but should not imply that literature based on research data and empirical social science would create a more *accurate* sense of place. Incidentally, the role of the reader does not change whether the text emanates from the author's senses or from qualitative interviews.

Sense of place is an essential element of the murder-mystery literary genre (Hausladen 1995: 63). It is crucial for storytelling and goes beyond providing the setting for the narrative (Wissmann and Zimmermann 2010). In a reference to the novels of Agatha Christie, Hausladen (1995: 64) comments that descriptions and creations of a certain mood in the text aim at generating sense of place.

> Description of settings emphasizes the natural and human landscape, its appearance, and its mood as affected by factors such as weather, time of day or season, and social interactions. Also, there are descriptions of events, as perceived by the storyteller. These descriptive elements provide vivid images of a specific setting (Hausladen 1995: 65).

As we can see, description in the murder-mystery novel is far more than setting the scene of the narrative; it is part of it. Sense of place not only exists in the literary genre but is of importance to the storyline and overall plot. In the end, it is the author's experience and skill that constructs a certain sense of place (see Gregory 1995: 30). Precise descriptions of the physical place help reveal an authentic sense of place – where authenticity is defined by strict analogy to possible real world experience (Hausladen 1995: 66).

> The characteristics of this genre and the relatively large readership it reaches guarantee that for certain purposes it is probably a more effective distributor of knowledge than are many kinds of serious literature (Hausladen 1995: 77).

Thus, media is identified as a creator of the sense of place. Readers might learn about places like Moscow only via media like murder mystery novels. Sense of place is created in an easy to digest way. As Kracauer (1979: 22) wrote about the detective novel:

> [U]sing generalized and simplified characters, places and situations, the novel provides society a clearer picture of itself than it would be perceived through their every-day experiences (Wissmann and Zimmermann 2010: 374).

Murphy (1996) radicalizes the influence of media on sense of place by describing it as the number one source for information about the world:

> People are exposed to the majority of current information, much of which is scientific and technological, through the mass media (Murphy 1996: 77).

A look at how perception works (see 1.1, especially Husserl 1983: 83) supports this statement it we take into account that most people – at least in developed countries – consume information mainly through mass media like daily newspapers and radio, books, and movies, TV and the Internet. In times of media and global communication, space, as well as time, has new meaning (Meyrowitz 1985: 13). The physical basis of communication no longer exists, which has also had an impact on place (see Massey 1991). Meyrowitz makes an interesting remark about the meaning of both place and sense. The latter, he writes, refers to "both perception and logic", while place signifies "both social position and physical location" (Meyrowitz 1985: 308). But social action is no longer bound to physical location. With or without physical place, social place is defined by information the individual is able to process. Husserl would call this "perception" (see Husserl 1999). Linked to communication and media, social behavior is ascribed to technological knowledge, and social behavior is not only shaped by mass media, but also influenced by it. TV and the Internet – as most influential forms – help to shape sense of place. (Chapter 5 will discuss the impact of media on the individual construction of the world using the example of audio drama.) The construction of sense of place through media shows that social roles are bound to social places and that – without a physical fundament – social roles become more arbitrary and even a loss of place might be felt (Meyrowitz 1985: 308). Thus, sense of place is described as potentially dynamic, at least in the long run. Meyrowitz also points out that the way information is transported and communicated changes our sense of place, not only in the present time but at all times. A more ancient example would be the printing press, which redefined place for a variety of social groups and roles:

> The sixteenth century also saw a striking change in the sense of place and in the degree of 'permeability' of the domestic sphere. Historians have noted the new sense of 'boundaries' dividing public and private, domestic and political, family and community in the transition between the pre-modern and modern age (Meyrowitz 1985: 314).

Paintings are another example of how media can shape places. Kennedy et al. (1988) describe the physical transformation of place through seventeenth-century landscape paintings: "[S]tylized English landscapes became the models for nineteenth century American city parks, private gardens, and valued landscape scenes, and continue to be followed today" (Kennedy, Sell, and Zube 1988: 32). The authors explore the deconstruction of the aesthetic meaning of a landscape and emphasize the impact aesthetic "attributes" have on place (Kennedy, Sell, and Zube 1988: 50).

As an example, Schnell (2000) describes the boundaries of the Kiowa Indians in their Oklahoma homeland:

> The region is important as a psychological anchor in the Kiowa past and as a core for their modern identity as a distinct people (Schnell 2000: 155).

"Homeland" is defined in a slightly different way than sense of place to explicitly focus on the social element of place attachment:

> What is a homeland, after all, if not a people's unique and unifying symbol of a common past and future? (Schnell 2000: 173).

The study of Schnell shows that sense of place is highly dependent on social context. It also points to the physical element of place that has been left out in many other studies (see Massey 1991). In the case of studying place attachment of the Kiowa Indians in Oklahoma, territory and locality become essential, even provisory. Without the physical entity of place, the concept of homeland would either not exist or have to be transformed to a dematerialized "home" that would only consist of common social negotiations.

Another example of using sense of place in relation to first-nation studies is the work of Windsor and McVey (2005). They describe the loss of sense of place for the Cheslatta T'En Canadian First Nation in Canada. Due to a large construction project, the Cheslatta T'En were "forcibly removed from their ancestral lands in the 1950s so as to allow for the construction of" a hydroelectric dam (Windsor and McVey 2005: 146). The relocation took place for of economic reasons and indifferent to Cheslatta T'En cultural heritage that, theoretically, must not have been obvious to the company that built the dam:

> A major reason for the situation with the Cheslatta rests with the differing 'sense of place' possessed by natives and non-natives (Windsor and McVey 2005: 158).

Whether this dispute over the Cheslatta T'En's sense of place was obvious to outsiders or not, this example shows how sense of place can be valuable in explaining and interpreting land-use issues by taking a geographical perspective and focusing on the social potential of place.

Sense of place is also being used in geographic gender studies. Christie investigates the Mexican house-lot garden as a place that "represent[s] people's symbolic connection with the land in increasingly urban contexts" (2004: 368). In a series of qualitative interviews Christie was able to connect with the women of various households to discuss food, kitchen, and the social potential of the house-lot garden. The garden represents not only Mexican history with a changing "relationship between society and the natural environment" (Christie 2004: 388) but also a place for women in a male-dominated household.

The relationship between women and men is also discussed in Valentine's "The Geography of Women's Fear" (1989). Here, the author doesn't focus on private but public places that are perceived as being dangerous by and for women (Valentine 1989: 385). Thus, "space" excludes certain locations, spots, and areas women avoid. Based on media reception, social narrative, and individual valuation, "[w]omen develop individual mental maps of places where they fear assaults" (Valentine 1989: 386). Time is an important factor for this concept of sense of place because certain environments only become hostile at certain hours, especially at night. The role of the media is also evaluated, describing victims as partly guilty because of their entering a zone dangerous for them. In other words, sense of place is used to justify the spatial limitations imposed on women. In the conclusion of her study, Valentine writes: "Women's inhibited use and occupation of public space is therefore a spatial expression of patriarchy" (Valentine 1989: 389).

Finally, Shamai and Ilatov (2005) try to systematize sense of place research. In a study about Israeli-born and immigrant groups they used a questionnaire to measure sense of place (Shamai and Ilatov 2005: 467). Place attachment is of primary importance to them when they ask, "what creates the sense of place, the perception of physical environment or the perception of the personal and social contact and interaction in the place" (Shamai and Ilatov 2005: 474). According to their results, quantitative data like occupancy, age, or sex are only partially accountable for the creation of a sense of place. In contrast to the media studies presented above, physical elements, such as concrete settlements, also matter for the establishment of a sense of place:

> Place is not merely a bounded area on a map; however, the physical attributes of the map are important too (Shamai and Ilatov 2005: 474).

In the end, empirical research about and even measurement of sense of place is possible. The quality of sense of place, however, seems to be highly uncertain. To get a better idea of the possible variables sense of place can contain, sound will now be highlighted within geographic sense of place studies.

1.3.2 Sense of Place and Sound

> Sensuous Geographies does not deal with the social issues of individual geographies, but does explore some of the dimensions of individual perceptions (Rodaway 1994: ix).

This quote from Rodaway shows that the physical senses of a human being are important when it comes to that human being's sense of place. One of them is the sense of hearing, which can be investigated from various perspectives, e.g., the consideration of blind and deaf people. For them, the dominant sense of seeing is contrasted with the loss or accentuation of hearing. In this context Rodaway points out that the "'quiet ear' may be silent but the body 'resonates', that is it can give at least a partial sense of the auditory world because sound is first and foremost vibration" (Rodaway 1994: 97). Sound is unable to provide an exact image of the environment because the mixture of sounds is always changing (Rodaway 1994: 104). Johansson and Bell highlight that music – Schafer (1994) would call it an organized form of sound – "is an integral part of the human experience" (Johansson and Bell 2009a: 1). Music is understood to be a carrier of cultural values that contains a multitude of social relations. Place identity and attachment can be created through music, political opinion is transported by it, and there is, of course, an economic aspect to music that impacts the outside view of place (Johansson and Bell 2009a: 2).

Matless (2005) examines the sonic environment of a wetland region in eastern England in relation to society's notion of "tranquil space". Sounds are investigated related to their belonging to a certain place: "[I]f sounds are deemed out of place this is not for intruding into a landscape lacking sound, but from disrupting a valued acoustic ecology already full of sound" (Matless 2005: 762). Tranquility does not have to correlate with silence but with sounds that go with the environment based on society's values. Sonic geography lays open the cultural elements, like aesthetics, of a region as it describes how the acoustic landscape – or soundscape – should sound and what kind of sounds are labeled as preferable, appropriate, or even solely right (Matless 2005: 747).

Sense of place in combination with sound is also part of studies about identity in human geography. Boland "examine[s] the extent to which sound, for example a distinctive accent and/or dialect, affects the construction of local identity" (2010: 1). The study looks at people born in Liverpool, England, and their distinctive Scouse dialect. What is interesting here is the attempt to define identity not by location but sound. Liverpool natives are generally named Scousers because of their vernacular. But language does not stop at city limits. And there might be a part of the urban population that does not speak Scouse and so would be falsely labeled Scousers.

> Can we accept a territorial, place-bound definition of Scouse identity? Is a sonic conceptualization equally or more appropriate? Put another way, is Scouse identity determined by where a person is born or how they sound? (Boland 2010: 6).

Thus, social identity is formed by sonically excluding or including certain people (Boland 2010: 17) – a continuation of Matless' study (2005) described above.

Coming back to an organized form of sound – and we will talk about different music genres shortly – culture and identity can also be expressed through music.

British-Asian cultural politics manifest themselves in dance music and musical performances in London (Jazeel 2005: 233). Britishness is constantly being renegotiated vis a vis English tradition, the cultural impact of former colonies, and new identities that arise out of the blend of the two. In this diverse context, music offers an opportunity to gain insights into how sense of place is being created (see Jazeel 2005: 239). This is not only true on a national level but also a local one, as the following remark from Lyshon et al. (1995: 430) shows:

> Rap and hip hop show not only the complex geography of musical formations but also the explicit engagement of popular music with issues of place identity.

Music might depend on where it is created, but it can also become the creator of place. It is most likely that there is a strong interconnection between the two. Different music genres were developed in specific places, mostly cities. Alternative rock, especially grunge, was born in Seattle – most popular bands were *Pearl Jam* and *Nirvana.* Techno was created in post-Fordist Detroit, respectively. Che (2009: 264) explains: "The sound of machines, part of the environment in Detroit, was also key to techno music". But in music, place can be (re)created as well: "Detroit techno has been described as hopeful, euphoric, and simultaneously anxious in its finding of 'beauty in decay' in the deindustrialized city" (Che 2009: 264). The special relationship between place, songs, texts, and covers has been described by several authors. Romig (2009) highlights economic conditions using the example of Hollywood's film industry and its need for musicians who produce soundtracks and theme songs. Living in Los Angeles or California as a whole has inspired musicians to work their thoughts on place into their lyrics. Place descriptions vary from band to band and change over time:

> Representations of California in popular music lyrics during the second half of the twentieth century changed quite radically from a distant, tropical, romantic paradise to a crime-ridden, drug-abusing misogynistic gang territory (Romig 2009: 119).

From the *Beach Boys* to the *Red Hot Chili Peppers*, California-based bands have evoked different places depending on their own historical backgrounds and imagination (see Daynes 2009).

> [T]he same city can be envisioned as two completely different places based on not only the lyrics, but also the sonic presentations (Pesses 2009: 147).

For musicians as well as listeners, music means more than just a harmonic sequence of sounds. It connects the individual to places and places to the individual, widening the experience of physical elements by internalized memories and current feelings (see Pesses 2011: 42). Thus, "the consumer is able to move beyond conventional notions of the body and its dependence on concrete locations" (Curti and Craine 2011: 137). That does not mean, of course, that the bodily experience is being denied: "In other words, sound, music, place and bodies are not completely culturally constructed" (Duffy and Waitt 2011: 121).

The mostly aesthetic connotation of a certain sense of place in music is backed by dissociation from places perceived as being negative (Said 2003, see Relph 1976). Using the example of Canada, Kuhlke (2009) points out that "national identity is often evoked through powerful metaphors of nature" but also "by creating a variety of imaginary musical landscapes that parody the United States and define Canadian society as diametrically opposed to America" (Kuhlke 2009: 162).

One globally exported cultural good of the United States is the music of *Elvis Presley*. Today, it is not only the pure music of "The King" or pop icons like *Michael Jackson* that is related to place. Aside from the interpretation of lyrics and tunes, there is a touristic aspect to the phenomenon that results in massive travel inside the fan community (Alderman 2009: 90). *Graceland* is more than a large mansion in Memphis, Tennessee. Because of music it has become a place for pilgrimage; it even became a National Historic Landmark in 2006. In their study of ambient music in Australia, Connell and Gibson argue that "[M]usic has contributed to the imaginative representation of a touristic" place (Connell and Gibson 2009: 67). Places can gain importance as touristic spots because to some tourists they represent physically what a certain genre of music communicates acoustically (see Connell and Gibson 2009: 83–4).

As music can attract tourists to visit a place, it also often derives its identity from where it was created (see Schyma 2002: 25):

> [M]usic is made in specific geographical, socio-economic and political contexts, and lyrics and styles are always likely to reflect the positions of writers and composers within there contexts (Connell and Gibson 2003: 90).

Hip-hop and rap, once used to signify the same type of music, express local culture; today they differ from each other not only by genre but space (Graves 2009: 245). In New York's Bronx borough, hip-hop expresses the physical and "social isolation" of its residents (Graves 2009: 252). Johansson and Bell point out that music genres don't have to be physically separated but can relate to the same place (Johansson and Bell 2009b: 220). There seem to be places where music is being produced more easily, where the social and physical environment inspires musicians. "[D]eindustrializing places in the UK in the late 1970s" are an example for the importance of place for punk (Johansson and Bell 2009b: 221). Climatic conditions, like the rain in Seattle, are often said to stimulate local music scenes. Cities with a harbor, college, or university often have a high influx of migrants who add to the creative potential of place (Johansson and Bell 2009b: 221–3). One social aspect of people living in a certain place is their faith. Sound, for instance religious music, may impact a whole city's media and event infrastructure. In Nashville, Tennessee, Christian concerts and music festivals are celebrated, religion-focused radio stations play mostly Christian songs, and religious music albums are sold in high numbers: "The very utterance of [...] [Contemporary Christian Music] posits the existence of a bounded and internally coherent set of Christian voters or consumers" (Lindenbaum 2009: 291). Regardless of the genre, music is a very emotional form of sound that impacts sense of place. Individual expression in music does not reside with the producers and artists only.

Bodily expressions and movement inspired by music are equally important.

> Musical definitions of Havana are at their richest and most personal at the neighborhood scale where shared cultural histories create unique soundscapes. At this scale the streets have rhythm as bodies, instruments, cars and buildings provide the structures that carry the beat (Finn and Lukinbeal 2009: 127).

The whole environment is part of the musical experience. In their study on Musical Cartographies, Finn and Lukinbeal point out the all-inclusive "affective reality of its creation and performance" (Finn and Lukinbeal 2009: 142). The organized form of sound (see Schafer 1994) appears to be the most intuitive one.

The means of distributing and accessing music also impacts sense of place. The Internet and traditional radio are the two major media players today. The Internet, as the latest innovation for promoting and consuming music, has changed not only how people listen to music but what they listen to. Kruse points out the changing "landscape of indie pop and rock music" (Kruse 2009: 205). Smaller labels that cannot afford big advertising and production budgets offer their catalogues on the same buying platforms the major labels do. Portals like iTunes are no longer reserved for global record companies. Global Internet distribution has had at least two effects: First, music is extending its reach so that more people can listen to local styles. Second, these local styles might get transformed from their local heritage toward more global meaning:

> And the Internet, with its ease of connecting people across localities, regions, countries, and continents, may well play a role in a decline in a sense of local identity and of being part of a particular local history within a music scene, and in the growth in the sense of translocal identity (Kruse 2009: 209).

But media can also be used to interconnect an existing community and strengthen bonds among its members. Especially for people living in diaspora, local music, globally available via radio, can help them stay attached to familiar places like their hometown or homeland. Despite this being the age of the Internet, Keough highlights radio as "the most important medium of mass-communication" (2009: 185). He describes music as "essential in [...] culture" (Keough 2011: 93). Listening to sounds from home preserves cultural heritage and belonging.

Finally, there is scientific potential for the sense of place and sound in the context of politics. If people listen to music on the radio, they might feel attached to their home and experience its sense of place, but music can also be exploited to transport political messages. Jackiewicz and Crane investigate music's potential to create "spaces of activism and resistance" (2009: 33). Music is understood as a cultural product that represents political and social opinion. Through its symbolic character, music describes the environment, creates borders, and overcomes political boundaries (Jackiewicz and Craine 2009: 46). Beyond its acoustic qualities, music's meaning can be used to transport political messages. For Kruse II, the "primary interest is not the Beatles music, but rather how they have influenced the meaning of places, place representation, and tourism" (2009: 11). John Lennon's political statements promoting peace in meetings with the press in Amsterdam and Montreal were only possible because of the potency of *The Beatles*' music.

As shown, geographic studies – at least to a certain extent – include music in sense of place research. To change the perspective from place to sound, the following section will turn away from geography and open the investigation to various disciplines in the humanities.

1.4 Sound in the Humanities

After discussing studies on sound in geography, this section will reveal different areas of sound research spread across the humanities and neighboring disciplines. Before we deal with the effects of sound on the urban dweller in Chapter 2 and acoustic ecology's methodological approaches in Chapter 3, an overview seems necessary.

Acoustic ecology's research from the 1970s has inspired authors to investigate soundscapes and the sonic environment from a phenomenological, ecological perspective (Atkinson 2007, Bijsterveld and van Dijck 2009, Blesser and Salter 2007). "The examination of sound begins with a phenomenology" (Ihde 2007: 17). Thus, a description of the urban environment includes sound in urban planning as well as architecture. Landscape values are related to sound (Carles, Barrio, and de Lucio 1999, Hedfors 2008), physical structures in the city are examined in regard to acoustic qualities (2010, Thompson 2004), and relations between architecture and music are explored (Young, Bancroft, and Sanderson 1993). Music is important to the perception of place in acoustic design, as Kastner (2008) points out, where sound can be contrasted to color and other visual impressions (Greated 2011). For informatics and GIS, visualization of sound in maps and models has become more and more important. Basic conditions for mapping are discussed since the end of the twentieth century (Kobayashi and Schmandt 1997, Krygier 1994, Servigne et al. 1999). Today, there are works that deal with 3D sound models of cities (Kurakula 2007) and whole countries (Edler and Lammert-Siepmann 2010).

Gunderlach (2007) answers the call of Schafer (1994) for a museum of lost sounds when he explores the acoustic features of historic places. "An Environmental History of Sound and Noise" is contributed by Coates (2005). Similar to historic studies on sound are ethnologic inquiries. Sound is used to explore the history and recent situation of places like New Orleans (Sakakeeny 2010). Social structure manifests itself in sound, and there are specific soundscapes ascribed to certain places (Feld 1984, Hahn 2002). These studies from musicology are closely related to sociology. Other works deal with the artistic qualities of sound (Fontana 1987), its constructivist potential to modify space (Jones 2006), and the aural border between cultures expressed through different music styles (Kun 2000).

Most studies about sound in the humanities are written in a sociological and psychological context. There is a wide variety of sound studies, especially within the discipline of sociology. The impact of sounds on humans is investigated by Díaz (2006), while other authors highlight the individually differing ratings of sound, including positive and negative connotation (Aglioti and Pazzaglia 2011, Kim and Rowland, Yu and Kang 2008, Yu and Kang 2009). How the world is heard individually is also subject to principles of psychology (Gaver 1993). Sound design (Visell et al. 2009), the effects of environmental noise (Kryter 2007), and motion (Hetherington, Daniel, and Brown 1993) are examined in relation to perception. The latter links sound studies to phenomenology again. Ihde reflects on:

> the history of philosophy with recollections of pages and pages devoted to the discussion of 'material objects' with their various qualities and on the 'world' of tables, desks, and chairs that inhabit so many philosophers' attentions: the realm of mute objects (2007: 50).

It is a simple but astonishing discovery that most philosophical inquiries look at objects in a non-moving, silent state. Chairs and tables are examined extensively, while movement – and with it sound – is often of less importance. But with sound, a "phenomenological approach [...] can include in the description of a city's attributes the corporeal, the sensual and psychological aspects of subjective experience, as well as the broader cultural characteristics of the different communities and subcultures which contribute to the diversity of city spaces" (Arkette 2004: 160). And as "[n]othing is more 'familiar' than our own experience, and nothing is closer to ourselves" (Ihde 2007: 17), the following studies will present this phenomenological approach – starting with acoustic ecology.

Whether sound is perceived as being painfully noisy or entertainingly loud is up to the listener. Places like live music venues have a certain impact on people. Sounds shape places; they even have the power to (trans)form the urban environment (Atkinson 2007: 1912). "A key example of the territorial control of commercial, and increasingly public, space can be heard in the functional music, or muzak, of many urban spaces" (Atkinson 2007: 1910). In section 2.3.1 we will discuss a post-Fordist challenge of sound (see Atkinson 2007: 1913): To actively affect listeners in their perception of and behavior in space.

> [P]eople, as listening bodies in urban spaces, [engage] both passively (via hearing) and actively (via listening) [...] with sound and unwanted noise (Atkinson 2011: 12).

In urban planning this affective potential of sound is often overlooked, as architects rely on an inherited visual perspective (Blesser and Salter 2007: 1). To help shape the perception of space and places, acoustic elements are included in the design process of buildings and other physical elements in the urban environment – labeled as "aural architecture" (Blesser and Salter 2007: 2). The surroundings of each person may be unique (Gibson 1986: 43), but the individually perceived soundscape always has an influence on place experience. Acoustic ecology has acknowledged that "the human body is a listening presence that both actively and unconsciously experiences sound" (Atkinson 2011: 13). Sound in the city is labeled "both an ordered and ordering force" (Blesser and Salter 2007: 24) that has more qualities than frequency and loudness. The social element is part of sound studies in acoustic ecology (Schryer 1998), as well as the different cultural meanings that go with it (Bijsterveld and van Dijck 2009: 12). Schafer's museum of lost sounds (1994) is still missing (Blesser and Salter 2007: 6), but there is an understood need to collect and restore past sounds (Bijsterveld and van Dijck 2009: 12).

> Industrialisation and new technologies have transformed our sound environment. New sounds have appeared, have transformed our concept of hearing and have modified our aesthetic criteria in a decisive fashion (Schryer 1998).

Based on individual tastes and social imprints, soundscapes are considered either positive or negative. Music, for example, has the potential to add to a place and thus shape the individual evaluation of it (Thomson 2005). Whether through the sound of shoes on the floor, cell-phone conversations, or the rustling of cloth, people actively add to the soundscape. Different sounds recur from time to time, but the single sound events of a certain baby crying or an individual person shouting are temporally unique. Sound is "the result of dynamic action" (Blesser and Salter 2007: 15).

Acoustic ecology identifies the man-made production of sounds as being mostly unconsciously produced. As Atkinson points out, in the early twenty-first-century quiet carriages are installed on UK trains because a cultural code calling for tranquil, quiet places wasn't being honored (Atkinson 2011: 19). Listening appears to be linked to the cultural background if certain people perceive the urban soundscape as being in perfect harmony. Schryer (1998) calls this The Sharawadji Effect, while others call it noise (see Atkinson 2007: 1915). In acoustic ecology, the phenomenological approach to understanding "how sound is experienced" (Atkinson 2011: 13) is described as follows:

> Hearing, together with its active complement, listening, is a means by which we sense the events of life, aurally visualize spatial geometry, propagate cultural symbols, stimulate emotions, communicate aural information, experience the movement of time, build social relationships, and retain a memory of experiences (Blesser and Salter 2007: 4).

Adding to this, Atkinson (2011: 14) highlights the "significant affective and behaviour-modifying influence" sound has. In the urban context, sound often derives from the physical structure of buildings. That is why sound is also being discussed in architecture.

"Does the acoustic environment correspond with the intentions behind the designed places, in general?" (Hedfors 2008: 11). Architecture is currently facing various problems related to sound and the urban soundscape. Whether in relation to the shape of a building, the materials used in its inside and outside facades, or the position of the structure in relation to the existing environment, sound matters. Therefore, questions about the building site relating to sound are critically discussed:

> From the very outset of any building development, the selection of the site, the location of buildings on the site, and even the arrangement of spaces within the building can, and often do, influence the extent of the acoustical problems involved (2010: 1).

Compared to theaters, concert halls, or smaller music live music venues, most problems arise with buildings that specifically do not need an excellent sound design. And if buildings contain a proper sound design, like noise insulation, the outside environment is often neglected in the construction process. It is taken as empty space, available to fill with new offices and homes, and seldom understood as being part of the overall soundscape. "Buildings become more airtight and soundtight while the outside environment of ever increasing walls and concrete is a declining haven for sound quality" (Dawson 1988: 170). But sound, as part of the urban environment, cannot be neglected (see Hedfors 2008: 15). Otherwise the city is confronted with acoustic pollution like traffic noise from nearby highways. After construction is done, the dissonance may only be solved by tuning out existing noise sources. Noise barriers, for example, are a symbol of giving up on acoustic design (Kleilein 2008: 7). Thus, some authors value acoustics neglected in the design process (Kleilein 2008: 6), while Newman, in the 1960s, wrote, "Acoustics is a concern in every single building project" (1960: 26).

In architecture, sound is also discussed directly in relation to place. Schafer's point of view still seems timely when nature sounds are described as valuable for the urban environment to enhance the acoustic setting of place.

> Many natural sounds (especially the sound of water) help to enhance both the images of natural environments and of urban spaces by projecting onto them a meaning other than that derived from the image in isolation (Carles, Barrio, and de Lucio 1999: 198).

In contrast, other authors relativize the impact of nature sounds and accentuate individuality instead: "Some find the amplified sounds pleasurable, as signs of the excitement of urban life; others find these sounds intolerable" (Dawson 1988: 170).

Overall, the more important attribute for place and sound is its appropriateness. "The 'right sound at the right place' enhances the character of the locations and highlights any adjustments which are made" (Hedfors 2008: 12) (see also Carles, Barrio, and de Lucio 1999: 200, Dawson 1988: 174). Aside acoustic ecology's focus on nature sounds, music in the urban environment is also investigated (Young, Bancroft, and Sanderson 1993: 39). Dawson's study on the acoustics of the urban garden calls for an integration of all local sounds:

> If environmental designers were sensitized to an acoustic profile for a community where individuality of neighborhood sounds set the pretext for gardens, then not only could new sound be added but historic sounds [...] might be enhanced (1988: 175).

Sometimes sound also provides space with some privacy. Aside from sound insulation, which has already been named above, adding sounds like water can, to a certain extent, mask conversations and tune out unwanted sounds – Newman (1960: 30) calls it "acoustic perfume". Whether sound or acoustic barriers are used to cover noise, architecture has to deal with it. In the beginning there are various noise issues that can always be solved, if architects attempt to do so (2010: 2). Afterwards, sound can explicitly be used to generate positive impressions of place. What works on the level of an inner-urban construction site can be adopted to city-scale or even statewide (Brakemeier 2010: 1). Sonic branding was first used in advertising, where products are often associated with certain sounds or a catchy melody (see Kastner 2008). With the translation of sonic branding from products to places, with the move to advertise places as products (Brakemeier 2010), visualization of sound becomes an issue.

Mapping sound is subject to many GIS studies and informatics that are challenged with visualizing something that originary cannot be seen. In urban planning, the starting point of most studies is the environment of urban places. Sound is understood to belong to place, where music, for example, is directly influenced by the environment (Adams 1994: 8). Sometimes, urban sounds still have a negative connotation, as referenced in Schafer's soundscape studies (1994): "In one hand, when music, sounds are considered as enhancing the quality of life, but in the other hand, traffic noise deteriorates the quality of life" (Servigne et al. 1999: 262). The overall aim is to highlight the existing sound situation of the city with its specific places (see Adams 1994) and to provide planners with a proper instrument to optimize the urban quality of life by showing how to banish all noise (Servigne et al. 1999: 263).

The variety of GIS and mapping projects involve sound in multiple ways. Sound can be used as a tool to enhance GIS, like with *earcons*, acoustic reminders of localized actions (Kurakula 2007: 150). Earcons can be embedded to enhance GIS browsing environments (Kobayashi and Schmandt 1997: 194), and upgrade cartographic e-learning material (Edler and Lammert-Siepmann 2010: 1). As a subject of research, two different types of visualization exist: 2D and 3D mapping.

"Most of the noise maps that are available today" (Kurakula 2007: i), are 2D, but "2D noise maps cannot provide enough information for decision making process and also for implementing noise mitigation measures" (Kurakula 2007: 16). As sound expands radially from its source, a 2D approach is problematic. High buildings and sound exposure in their upper levels cannot be displayed properly. On the other hand, grey scale, color, and line symbols provide an easy-to-read overview of existing sounds in a city. Light and dark textures of sound can be illustrated by contrasting white, grey, and black areas. Warm and cold sounds can be represented by red and blue, respectively. Isophones, lines of similar loudness, are another option for visualizing sound in 2D mapping (Servigne et al. 1999: 265).

3D models are often more effective.

> Abstracted two-dimensional space and the visual variables – the traditional purview of cartography – may not always be adequate for meeting the visualization needs of geographers and other researchers interested in complex dynamic and multivariate phenomena (Krygier 1994: 149).

To display sound that expands radially (see Kurakula 2007: i) and to consider multiple variables, GIS has to use 3D models. Where 2D visualization stops, 3D versions are able to cover multiple layer of information. Weather conditions, traffic flow and speed, even pavement and day and night settings can be displayed (Kurakula 2007: 2). Coupled with the visualization of sound and especially noise also caused by the urban population – like high rise buildings and high traffic volume (Kurakula 2007: 1) – the impact of sound on the urban population can be displayed in reverse.

Urban sound research in ethnology is not about maps but individual stories and social settings. Although studies and journal articles investigate rhythms, beats, and music as a type of cultural representation, "very few of them include sound or the option to connect to a website to hear the sounds that are part of an argument or interpretation in an essay" (Feld and Brenneis 2004: 470). Authors like Sakakeeny focus on the importance of sound for the social construction of place (2010: 3). Examples like New Orleans jazz funerals show how music in certain settings creates place and transforms a neutral space under an interstate bridge to a spot with social meaning:

> The 'bridge' creates intimacy, enclosing parade participants, maximizing a sense of unity, and the concrete makes for spectacular acoustics, amplifying and multiplying the participatory sound, creating a sort of 'unplugged' feedback loop (Sakakeeny 2010: 2–3).

The argument, that individual openness to sound determines how a person experiences a soundscape (Sakakeeny 2010: 4), reflects the mind-set of historians who work on sound. "Changing attitudes to sound are as important as the changing nature of sounds themselves" (Coates 2005: 641). That comment describes a problem for historical science. Sound is fleeting and, thus, hard to preserve (Gunderlach 2007: 13). The timeframe for most historical studies on art installations, environmental sounds, and environmental speech – the latter as part of environmental sounds (Coates 2005: 638) – is the turn of the nineteenth century.

The contemporary artistic use of sound is explored in the fine arts. The importance of sound differs as it can add to predominantly visual installations or be an equal partner to the visible (Jones 2006: 22). Greated calls those works "most rewarding [...] where sound and vision go hand in hand" (2011: 347). Indeed, most sound art includes visual elements, at least as part of the installation. Visible elements are also considered unfavorable for sound art:

> Visible speakers would have greatly weakened the psychological impact of the sound sculpture (Fontana 1987: 146).

But even a pure sound installation is located somewhere. Environmental surroundings have in impact on how a work of art is perceived. Sound "is entering the visual arena" (Greated 2011: 345), but the artist can use sound to alter the experience (Greated 2011: 343). An example by Fontana (1987: 144) illustrates the point: The place where the ruins of the former train station Anhalter Bahnhof in Berlin, Germany, are located was acoustically enriched with sound from current trains. The combination of sound and visual remains created a new environmental experience and, thus, a new environment. Studies in musicology link to many fields of research, like the above-mentioned art, ethnology, and history. Hahn's collection of sounds artistically makes hearable the soundscape of Zagreb, Croatia (2002). Feld's inquiry into the musical lives of the Kaluli people in Papua New Guinea is ethnologically bound (1984). Historically, musicology has examined sound waves as the descendant of musical notation (Filimowicz and Stockholm 2010). Kun combines artistic, ethnographic, and history approaches when he broaches the issue of the Aural Border, the acoustic barrier between the USA and Mexico, manifested in rock music originating in the United States but sung in Spanish (Kun 2000: 16). Here, music creates an "Audiotopia" (Kun 2000: 6), an acoustic place of difference, founded in territorial characteristics, cultural divergence, and local significance.

A phenomenological approach in musicology comes from Filimowicz and Stockholm (Filimowicz and Stockholm 2010). Here, the visual element is again of importance to the deconstruction of the meaning of sound.

> When one hears a door slam, one knows that the sound came from the door. When one hears a computer-generated waveform, one may know that the sound came from a computer, or from electronic circuits, or from amplified speakers. But the sound does not issue from these sources in the same way that sound issues from [...] a slammed door. There is a certain crisis in connotation, a failure of reference (Filimowicz and Stockholm 2010: 5).

Even without knowing the source – and this is a fascinating thought – sound is always associated with an image.

Images based on sound, or "sound images", not only exist for single objects like a door being slammed, but for the city, too (Fortuna 2001: 1). In sociology, sound experience is based on individual perception and social interaction. Classical piano works represent Western culture; its sound transmits cultural values (Baier 2009: 7). There are authors who tend to derive certain sounds, and not their interpretation, from the cultural background – i.e., values, labour, luxury, etc. – of the individual or group. Concepts like globalization are understood to be hearable in relation to world or global cities (Meier-Dallach 2007: 1). Urban sound itself is deconstructed and described in detail: "The predominant sound of the city is now a mechanical, rhythmic sound with a regular, continuous, and routine cadence" (Fortuna 2001: 10). The sound of the city is a mixture of single sound events from traffic, music, conversations, and other sources, that represent decline of social relationships:

> The overlapping sounds that constitute the soundscape of a public office or of the informal atmosphere of the city street are not identifiable, single, individual sounds. [...] They symbolise an urbanity made of mixtures of sounds and of the loss of relationships of mutual knowledge (Fortuna 2001: 11).

Using the example of Madrid, Spain, Díaz and Pedrero (2006: 271) discuss the effects of being exposed to the variety of urban sounds. Yu and Kang go a step further by exploring the "effects of social, demographical and behavioral factors" on how the urban dweller experiences sounds in the city (2008: 772). In a subsequent study the authors ask about people's expectations of the urban soundscape in certain public locations (Yu and Kang 2009: 1163). This individual's experience of sound is contrasted by comments that understand sound – in this case in its negative sense as noise – as independent from "social and cultural differences" (Hahn, Pethes, and Stäheli 2003: 206). Indeed, noise is described as being socially constructed – at least the definition of what noise is – but its overall impact applies to everyone. One possibility for avoiding sounds like noise is to create an individual soundscape for the surrounding environment.

> Does the typical individualism of the urban world have a specific sound? The answer is yes if we try to ascribe meaning to the sounds of devices like the beeper, the mobile phone and the Walkman. All of them are technologic artifacts designed to filter communication, aggressively exposing individuality (Fortuna 2001: 12).

The individual experience of sound always has an influence on people's behavior (Kim and Rowland: 3). Other factors, like vision and olfaction, add to the apparition of places, but hearing can sometimes become the dominating sense (Aglioti and Pazzaglia 2011: 47). Places can be imagined through acoustic stimuli but they have to have a cultural meaning in order to impact people's behavior and perception. Sound events like the ringing of a church bell are the acoustic expression of the church as a cultural artifact (Kim and Rowland: 6). Thus, sound "operates within place" (Kim and Rowland: 8) to shape it and evoke a certain image and understanding.

Psychologically, sound has a huge impact on the experience of the environment. In a study about media representation of an everyday environment, respondents valued dynamic surrogates that included sound in addition to static photographs more similar to the original (Hetherington, Daniel, and Brown 1993: 290). Sound is part of the dynamic environment, and single events are remembered quite accurately. People tend to almost exactly recall tempi of certain pieces of music, and even environmental sounds' BPM (beats per minute) can be described accurately (Boltz 2010: 168).

> Many sounds have a characteristic tempo that remains relatively invariant over successive occasions which thereby enables one to recognize the identity of the particular individual/object producing the sound (Boltz 2010: 178).

Interpreting walking sounds, for example, can be important for social interaction (Visell et al. 2009: 947). Rate memory of a friend's walk, in this case, is useful for predicting estimated arrival time (Boltz 2010: 178).

In terms of hearing, there is a difference between everyday listening and musical listening: "[I]t is possible to hear any sound in terms of its source (everyday listening) or in terms of its sensory qualities (musical listening)" (Gaver 1993: 286). As shown above, hearing a sound always evokes an image (see Filimowicz and Stockholm 2010), so that the individual descriptions of sound events tell about the social, cultural, and individual background of the respondent. That the individual stream of consciousness (see Husserl 1997: 167) not only helps to shape sound (experience) but is also affected by it will be discussed in the following chapter.

Chapter 2
Sound Effects

2.1 Various Effects of Sound on the Urban Dweller

Consciousness and cognition are both affected by sound. Aside from seeing, the sense of hearing is most important when it comes to understanding the world we live in. Before the possible effects of sound on the listener and her/his stream of consciousness are discussed, the physical process of hearing will be illustrated in section 2.1.2. As several studies show, the physiognomy of the ear is responsible for how listeners hear (Byrne, Michael, and Tufts 2011, Clark and Cox 2012, Irwin et al. 2011, Newman 1960, Nilo 1969, Rodaway 1994). A description of the outer, middle, and inner ear will be given, including neuronal processes that enable listeners to experience the environment acoustically. Harmful effects to the human ear can disturb hearing both temporarily and permanently, depending on several sound qualities like the sound level (Roizen 2003). Non-auditory effects – especially of unwanted sounds – will be discussed afterwards (2.2). If unpleasant sounds and noise do not permit listening, acoustic signals reaching the ear have to be processed and translated into meaning. Aside from the physical ability of hearing, the cultural setting is of importance, when sounds like spoken language have to be understood *right*. What is being heard always has to be interpreted based on individual significance, social meaning, and cultural context (Gaver 1993, Gerry, Faux, and Trainor 2010).

Section 2.1.2 emphasizes the importance of sound for perceiving the world. The urban environment is full of sounds, whether we listen to them or not. Sound is part of everyday life and thus always affects us (Irwin et al. 2011, Newman 1960). Media like TV or film sound completes the picture of the world, which can also emerge from sound-only audio media (Kim and Iwamiya 2008). Music as movie soundtracks, audio drama, or as stand-alone production affects how the world appears to us (see Chapter 5). Music is also suitable for expressing ourselves through performance or consumption (Morley and Somdahl-Sands 2011). Language and speech signify another effect of sound: communication. Acoustic stimuli are processed inside the human brain to enable us to talk to each other (Fennell, Byers-Heinlein, and Werker 2007, Nilo 1969). Even before proper articulation of a language is learned, infants use different intonations and sounds to communicate with their parents (Wood and Gustafson 2001). On the one hand, sound can even have a therapeutic effect on humans. Pain, for example, can partly be masked by music, and positive association with a certain sounds can have a promoting effect on future health care (Kristjánsdóttir and Kristjánsdóttir 2011). On the other hand, noise can be a stressor that negatively affects the listener. Permanent, noise-induced hearing loss can be one of the worst results, while annoyance can vary from light disfavor to sleep disturbance (van Kempen et al. 2010). After highlighting possible effects, the following sections (2.4 and 2.5) will review in more detail both the negative and positive effects of silence and sound.

2.1.1 Hearing

"What in the World Do We Hear?" is the title of one paper dealing with auditory perception (Gaver 1993). The question could also be *how do we hear*, to focus more on the listener as a receiver than the source of sound. In a constructivist understanding, what we hear depends only on the processing brain of the listener. Phenomenologically speaking, without the source there would be nothing to listen to. Sound emanates from its source omnidirectionally, if there are no barriers like buildings, of course. The farther from the source, the more quiet the sound, reducing "at a rate of 6 dB per doubling of distance" (Cavanaugh, Tocci, and Wilkes 2010: 134). Whether a sound can be heard depends not only on distance from the source but also on the surrounding soundscape, and the individual capabilities of the receiver, i.e., the listener. The older the listener, the worse and less accurate s/he can hear (Gerry, Faux, and Trainor 2010: 95). In a quiet environment, sounds are better heard than within a soundscape that contains other, masking acoustic impulses. "In the stillness of night, it seems we hear more distant sounds than in the bustle and heat of day" (Gerry, Faux, and Trainor 2010: 94). Architecture that considers spatial acoustics can help to enhance listening (Cavanaugh, Tocci, and Wilkes 2010: 134). It is complicated to include acoustics in the design of open spaces, buildings, and rooms, which is primarily focused on the visual outcome.

In contrast to seeing, hearing is multidirectional (Gerry, Faux, and Trainor 2010: 92).

> The ultimate evaluation of acoustical quality lies with the people who listen, speak, play, work, create, live, sleep, and otherwise use the rooms [...] This evaluation will differ from person to person [...] (Cavanaugh, Tocci, and Wilkes 2010: 211).

In general, human ears "respond to a remarkably wide frequency range that covers approximately 20–20,000 Hz" (Byrne, Michael, and Tufts 2011: 1517) – i.e., 20 Hz to 20 kHz (Clark and Cox 2012). Some authors widen the range for young people by putting the lower threshold at 16 Hz (Blauert 2001: 2, Rodaway 1994: 93). Anything below 20 Hz – in some cases 16 Hz – cannot be heard by humans and is described as *infrasound*. On the other end of the scale, inaudible frequencies begin around 20 kHz and are classified as *ultrasound* (Clark and Cox 2012: 81). Without discrimination against low frequencies humans would hear the sounds of their own body, "such as the movement of blood in our veins" (Rodaway 1994: 93). As hearing differs depending on age and physical condition, sound between "threshold and maximum tolerable level is called the dynamic range of the ear" (Clark and Cox 2012: 81). This range covers such a large number of frequencies and pressure variations of the sound level whose single values are indicated by a logarithmic decibel scale.

The human ear detects auditory information emanating from a source. The eardrum vibrates according to the acoustic stimulus and thus differentiates between frequencies.

> The ear is a complex organ dedicated to collecting vibrations from the air and transforming them into nerve impulses which are interpreted by the brain [...] (Rodaway 1994: 91).

Clark and Cox give a good analogy to describe the wide range of frequencies the ear is able to detect: "An eardrum in St. Louis that moves back and forth 1 ft at threshold, would move from St. Louis to San Francisco to New York to St. Louis for a very loud sound" (Clark and Cox 2012: 81), about 1,894 miles. The eardrum is located between the outer, or external, and middle ears. Sound reaches it through the ear canal, starting at the auricle, colloquially named the *ear*.

> Sound waves impinging on the eardrum cause it to vibrate, and these vibrations are transmitted to the fluid-filled inner ear (Byrne, Michael, and Tufts 2011: 1518).

> Stimulation of these hair cells produces electrical signals corresponding to the sound vibrations that entered the ear. The electrical signals are then transformed into their neural counterpart in the auditory nerve which is attached to the inner ear. The neural impulses are transmitted via a neural network to the brain for interpretation into meaningful sound (Nilo 1969: 114).

Sound is not always processed in the same way, nor even in the same parts of the human brain. The individual stream of consciousness of the listener co-determines how sound is interpreted. Irwin et al. allude to the fact that sound generally stimulates in two different ways. "Brain regions in the ascending auditory system [...] produced a significant response to the acoustic aspects of the stimuli" (2011: 266). But not only sound per se, with its frequency and level, is transmitted through the ear. What vibrations of the eardrum cannot translate into neural impulses is the associated emotion that goes with the things heard. But "brain regions in the paralimbic system [...] were engaged by the emotional content of the sounds" (Irwin et al. 2011: 266). So sound has two connected, yet separate, effects on the human brain related to acoustics and emotions.

A sound level that is too high results in temporary or even permanent hearing loss. Acoustics can amplify loudness (see Newman 1960), but whenever the source emanates sound above a certain decibel level, human ears are harmed. As a result of sound exposure of about 75 dB(A) during a whole day the "ears can become fatigued" and the threshold can change so that very quiet sounds are temporarily inaudible (Clark and Cox 2012: 84). Exposure to sound levels from 85 dB(A) to 90 dB(A) for a relatively short period of time – e.g., music in a discotheque for a few hours or walking by the turbine of an aircraft for only a few seconds – also results is a temporary decline in hearing (Rodaway 1994: 95, see Stansfeld and Matheson 2003: 243). Prolonged exposure to this sound level – e.g., in a factory – can even cause permanent hearing loss. "This type of injury is called noise-induced hearing loss" (Clark and Cox 2012: 84). The hair cells "located in the cochlea" of the ear are damaged and cannot be repaired "by medical or surgical procedures" (Byrne, Michael, and Tufts 2011: 1521). Especially low sound levels of 80 dB(A) to 90 dB(A) are not always perceived as being harmful. Children may not even notice hearing loss (Nilo 1969: 114). In case of acoustic trauma the sound pressure level generally exceeds 140 dB(A), causing immediate pain and permanent damage. The "inner ear tissues [are stretched] beyond their elastic limits, and [the acoustic energy can] rip or tear them apart" (Clark and Cox 2012: 84). In addition to noise and high sound pressure levels, various diseases can be responsible for hearing loss: hypoxia, hyperbilirubinemia, meningitis, congenital infections, ototoxic medications and substances, head trauma, otitis media, hypothyroidism, and other unusual etiologies (see Roizen 2003).

What we hear depends on our individual sense of hearing and is limited to the frequencies and sound levels the human ear is able to process. But there are cultural constraints too. Cultural setting influences what we hear, starting with the familiarity of the mother tongue (see Waldenfels 1999: 184, Wissmann 2011: 134). Like the varying stimulation of different brain parts, we can listen to the acoustic quality of sound – i.e., everyday listening – or "we might be concerned with the patterns of sensation the sounds evoke ([i.e.,] musical listening)" (Gaver 1993: 1). *Kindermusik* is one element of an acoustic cultural code. Music and movement are combined, and the "participation of the child" is the most important element in learning culturally dominant rhythms (Gerry, Faux, and Trainor 2010: 546). Songs about *the wheels on the bus [that] go round and round all through the town* ask for participation and foster a duple time beat that is easy to reproduce for Western listeners (see Gerry, Faux, and Trainor 2010: 550). Social and cultural preferences are easily learned and co-determine how and what we hear.

The experience of sound is not bound to acoustics and the human ear. "The ear forms the main sense organ in this perceptual system but, […] we can *hear* with more than our ears and the context, or environment itself, plays a key role in what or how we hear" (Rodaway 1994: 84). Accordingly, the possible effects of sound concern not only the human ear but the whole listener. Those effects will be investigated in the following sections.

2.1.2 Possible Effects of Sound

> Everywhere we go we hear somebody saying, 'I wish we could do something about the acoustics'. 'We hear our neighbors too much'. 'We can't hear the

> preacher in our church' 'The school cafeteria is too noisy; why don't the kids be quiet?' (Newman 1960: 26).

There are various situations in everyday life where sound affects the listener. The above examples seem to illustrate that sound is mostly perceived when it bothers somebody, while we pay far less attention to wanted sounds (see 2.3). But from a phenomenological perspective the question indicates something more basic: that sound is a vehicle of communication between the listener and her/his world. What sounds appear to be annoying or not depends on cultural coding. As discussed before, from childhood onwards people develop a "culture-specific metrical perception" (Gerry, Faux, and Trainor 2010: 545). In addition to the meaning of sound, "[t]he auditory system is highly responsive to sound level" (Irwin et al. 2011: 258). The experience of the urban environment – generally the loudest place for people to live – differs from respondent to respondent. As the following sections will show, sounds that one listener might find annoying can be joyful, even relaxing, for another person – loud music and the sound of a motorbike engine are typical examples. The visual element of the sound source has to be considered, too. The interplay of sound and sight result in an individual ascription of a sound as wanted or unwanted.

The potential to evoke a certain mood is not exclusive to environmental sound only. Sound not only offers a way to connect the listener to her/his environment, but to open up totally new worlds. As the example of popular culture audio drama will show, ambient sound, music, and language are used to imagine the world (see 5.1). "[I]n film and television, music is used to impart the circumstances and mood of scenes" (Kim and Iwamiya 2008: 429). Pictures without sound do not work in a movie. Something is missing. In the case of non-visual media, however, like audio drama, radio, or audio podcasts, sound is the only way for the listener to learn about the world. The images evoked in the production might even influence the perception of the listener's environment – which will be discussed in 5.2.

The sound of TV and movies, of audio drama and radio, and of music both live and recorded not only imagines the world and shapes perception, but is used to express individual feelings as well. Producing, performing, and also consuming music triggers emotions and fosters particular behavior. The consumption of music has various effects. Forsyth et al. believe there is a "possible relationship between drug use and music style" (1997: 1317). This does not mean that listening to a certain music genre promotes drug use, but that there seems to be an analogy between groups who take drugs and those who listen to music. As an example, the authors investigate rave music: "Rave music was positively related to the use of all categories of legal and illegal drugs, not just 'dance drugs'" (Forsyth, Barnard, and McKeganey 1997: 1324). Maybe rave music as a genre goes well with drug use, maybe not. But both music and drugs are associated more with emotions than logical reason. Sometimes, music can have a drug-like effect itself. Many studies describe music as a dominant, even crucial element of rituals. Ecstasy, rapture, and states of trance are accompanied by music that helps to connect devotees to spirits, gods, and other supernatural beings (Becker 1994, Brazeal 2003, Finney 1947, Jackson 1968, Jankowsky 2007, Rouget 1985). Section 2.3.2 will discuss this in more detail.

Beliefs can also be affected by music. "Music, both instrumental and lyrical, has the ability to foster desirable social and political attitudes and behaviors" (Morley and Somdahl-Sands 2011: 59). Most of the time, the entertainment factor seems to be more important than the political message of the artist for the listener who intentionally turns on a special song or album. During a commute, the car is used as an extended living room with hi-fi qualities (see section 2.5, especially Bull 2001a). But at least with live music it seems to be different, where entertainment is at least accompanied by another sound effect:

> [A]s we select our personal musical preferences and pay for entrance to a stadium concert, we are taking an implicit stand regarding the competing values of taste espoused by the artists we support (Morley and Somdahl-Sands 2011: 59).

This positioning of the listener, including identification with the artist and music, can be found in genre preference or individual – e.g. sexual – sympathies, but political and social positions can also be expressed. Artists like *U2* support critical thinking, not only in their songs, but also by using their prominent position to make political statements. "It is through their music and concerts that *U2* is able to influence their audience's perceptions and foster social change" (Morley and Somdahl-Sands 2011: 71).

Through the sound of music the artist communicates with the listener. But every listener is also a producer of sound, which is the basic requirement for acoustic interaction. "Of all the forms of communication, use of the spoken word is basic: it is a two-way process of speaking and listening, and hearing is critical to both" (Nilo 1969: 114). The effect of sound to enable us to communicate cannot be denied. But individual interpretation of different sounds is also crucial to ensure communication via language. Hearing and listening, in this case, cannot be separated. The significance of the mother tongue has already been discussed. Different sounds have different meanings and effects that can only be differentiated if the listener is embedded in the related social and cultural context. "The use of native-language phonetic categories to guide word learning is a crucial step linking perceptual development to language acquisition" (Fennell, Byers-Heinlein, and Werker 2007: 1511). Without understanding a language's phonetic categories, a listener could not even recognize single words. This structural, almost radical, alienness (Waldenfels 1997: 36) is overcome in the process of language learning, beginning in infancy. Eventually phonetic details like the difference between the letters *d* and *b* can be identified, leading to a better understanding – and much stronger impact and effect – of a language. Whenever an acoustic stimulus of a phonetic detail or category is heard, it has to be computed, set into context, and be filled with meaning. The latter is closely bound to the social and cultural context of the listener. "Through hearing, the infant perceives the speech patterns of others, and through hearing, he monitors his imitative speech attempts" (Nilo 1969: 114). Contextual interpretation of sound starts with infants' pre-linguistic sounds.

> [T]he cry sound does indeed vary with the infant's physiological dis- tress and, furthermore, [...] adults are sensitive to these variations or gradations (Wood and Gustafson 2001: 1287).

Successful acoustic communication, in this case, leads to an improvement of the child's situation.

Feeling better as a result of hearing sound is an effect I will discuss in detail in section 2.3. Compliments and endorsements are examples of the positive influence of language. Music can lift a listener's mood by stimulating certain parts of the brain, causing a feeling of happiness. It is also used to mask pain and unwanted associations with certain actions. Kristjánsdóttir and Kristjánsdóttir discovered that music provides pain relief during vaccination. In an experiment "adolescents receiving musical distraction were less likely to report pain compared to the control group" (2011: 19). The authors believe that music, when used during vaccination, can have a positive effect not only on the outcome of a single consultation, but on future occasions, too: "Building a positive experience around immunization thus can lead to less fear and more positive health care experiences in adulthood" (Kristjánsdóttir and Kristjánsdóttir 2011: 19).

There are also numerous negative impacts sound can have on the human body (see also section 2.2). Non-auditory effects are caused by both the acoustic and emotional qualities of sound, but go beyond harming the sense of hearing. Inter alia, symptoms can be "headache, vomiting, stomachache, difficulty falling asleep, or [...] [feeling] sleepy during the day" (van Kempen et al. 2010: 2819). Annoyance due to heavy traffic, live music, and air traffic noise promotes those harmful effects. As discussed above, very loud sounds directly affect the sense of hearing. Temporary and permanent hearing loss can be the result, mostly occurring in an urban or urbanized environment. "Noise-induced hearing loss [NIHL] or noise-induced hearing threshold shifts (NITS) can be a progressive problem" (Roizen 2003: 125). Both conditions usually aggravate over the years but do not make "someone unable to work" (Thurston 2013: 374).

> When people first became aware that loud sounds could cause permanent damage to their hearing is not known for certain, but one might expect that by the time metals were being forged, the metal workers would realize that the noise produced by their hammering caused changes in their hearing (Thurston 2013: 369).

Acoustic trauma, accompanied by permanent hearing loss, due to exposure to a sound level above 140 dB(A) is another harmful effect, directly affecting the sense of hearing (see Clark and Cox 2012: 84).

2.2 Unwanted Sounds

The effects of unwanted sounds vary from annoyance to severe side effects like cardiovascular problems and permanent hearing loss. Unwanted sounds can be those like operated machinery or an expression of individual dislike of certain sound events, like a voice in a conversation. They can also be actively caused to harm the listener, provoking pain as a method of systematic torture.

Section 2.2.1 deals with noise as a stressor (Smith 1991, van Kempen et al. 2010) that causes various annoyances and cannot be displayed by sound level only (Adams et al. 2006). Individual and social aspects play an enormous role in how sound is evaluated, as the following examples illustrate: A negative attitude toward the visual emitter can impact the evaluation of its sound (Pedersen and Larsman 2008), just as its color might negatively impact perception (Fastl 2004). Long-term exposure to noise lets listeners adapt to unwanted sounds and thus rate them more positive (Kryter 1972). Generally, sounds like those from air traffic (see 4.2.3) always have an individual impact on each affected person (Lavandier et al. 2011). The evaluation of sound is also affected by social processes (Maris et al. 2007), as the example of community noise shows. It "can be characterized as being perceived as an aversive component of a setting" (Jones, Chapman, and Auburn 1981). As Alam et al. (2010) point out, noise sources like food centers, children's playgrounds, and sports fields especially have an impact on high-rise dwellers. Group dislike for such types of noise can be described as community annoyance (Lim et al. 2008). They also discuss where unwanted sounds generally come from (Salomons and Berghauser Pont 2012), and several outdoor and indoor sources are identified (Banerjee and Southworth 1990, Ingerslev 1972). The negative effect of noise to mask wanted sounds like speech in a conversation (Ng 2000) is contrasted with possibilities to tune out unwanted sounds by adding positively connoted ones like water or other nature sounds (see 3.1) to the soundscape (Coensel, Vanwetswinkel, and Botteldooren 2011). They focus on the secondary impacts of unwanted sounds by providing a detailed description of the non-auditory effects of sound. Music consumption, for example, might reduce performance at work (Smith and Broadbent 1991), distract and thus increase reaction time (Tassi et al. 2012), and safety might drop if warning signals cannot be heard (Byrne, Michael, and Tufts 2011). Among the most common non-auditory effects described are sleep disturbance (Kluizenaar et al. 2011), cardiovascular diseases and endocrine responses (Institute for Environment and Health 1997), annoyance (Stansfeld and Matheson 2003), and disrupted communication (Greater London Authority 2004). Special age groups like children are sometimes singled out to better describe disturbing effects (Öhrström et al. 2006, Stansfeld et al. 2009).

Section 2.2.2 starts displaying lighter forms of unwanted sounds. The threshold of hearing loss (Stansfeld and Matheson 2003) and tinnitus effects are discussed (Hobson, Chisholm, and El Refaie 2010). To illustrate that sound always had the potential to negatively influence the listener, Johnson and Cloonan (2009) describe the tempting songs of the Sirens of the *Odyssey* and the chanting of the German *Lorelei*. Sound was used to demoralize detainees in Nazi concentration camps, while the *Nueva Canción* illustrates the power of music to frighten political enemies (Neustadt 2004). Today, so called *Mosquito* devices are installed in public places like playgrounds to displace lingering teenagers by emitting a painful sound only audible to this age group (Holmes 2012, Parliamentary Assembly 2010). Acoustic torture techniques in detention camps like the one operated by the US in Guantánamo Bay, Cuba, employ the most extreme unwanted sounds. Detainees are exposed to very loud noise for up to several weeks, even months, to prevent them from sleeping and to ensure total obedience to the torturers (Committee on Armed Services 2008, International Committee of the Red Cross 2007).

2.2.1 Annoyance

Unwanted sounds cause responses in the listener that do not have to be negative initially. The body reacts to the acoustic input of noise – e.g., music that is too loud or voices that are too aggressive – with a certain physical arousal:

> The arousal responses of the body are normally useful in that they increase the alertness of the receptor, neural, and skeletal muscular systems that would be involved in the organism reacting to the environment in ways that would protect it from potentially harmful events (Kryter 1972: 389).

In addition, certain sounds act as a stressor leading to secondary health threats (van Kempen et al. 2010: 2826). Triggers for both arousal and stress responses to unwanted sound – like fear – have more than just an acoustic quality. Whether a certain physical or psychic reaction is caused by sound or, for example, a visual impression, is hard to determine (Kryter 1972: 389). Acoustic stimulus is defined not only by sound level. Unwanted sound has a lot to do with the individual and social context of the listener and the surrounding soundscape of the source.

> [D]istance, temporariness and control over noise, are [...] [also] relevant to whether people would want to see a particular sound eliminated from their soundscape (Adams et al. 2006: 2385).

These elements of unwanted sounds lead to pollution of the acoustic environment. Unlike air pollution, noise pollution is a local phenomenon, caused by local sources and affecting only listeners within hearing range (Adams et al. 2006: 2396). In the urban environment, where “noise is increasingly seen as a key quality of life issue” (Alam et al. 2010: 3511), there is more than one sound source that is responsible for noise pollution.

> It is important not to consider issues of road traffic noise, aircraft noise or industrial noise in isolation from the tapestry of processes that together create the cacophony [...] of the city (Adams et al. 2006: 2394).

What sources cause unwanted sounds is highly dependent on individual perception and social context. First of all, a sound has to be actively listened to or at least heard to have any effect on the listener. "When participants do not listen to the sound and pay attention to the task, the tonality has no influence on activity disturbance" (Lavandier et al. 2011: 175). If sound is experienced, perception may not match common certainties, as "there is a disparity between [...] noise policy [...] and the way in which people respond to their soundscape more subjectively" (Adams et al. 2006: 2394). Other stimuli like color can influence sound-quality evaluation. In an experiment Fastl (2004) discovered that objects colored red were thought to be significantly louder than those of green color. Different sound sources have different impacts on listeners, who themselves all experience sound differently.

> A negative visual attitude [...] enhanced the risk for noise annoyance and possibly also prevented psychophysiological restoration possibilities. Aesthetic evaluations of the noise source should be taken into account when exploring response to environmental noise (Pedersen and Larsman 2008: 379).

Even if the assumption that "[e]xtroverts tend to have higher sensory thresholds" (Jones, Chapman, and Auburn 1981: 45) is questionable, individual and social aspects always have to be considered (Maris et al. 2007: 3493). Sounds like the human voice shouting or crying are not unwanted or annoying per se. Conversations and music can both be experienced as wanted or unwanted. Why a sound is experienced as unwanted can be linked to a personal fear of the source of it. Mosquitos, sirens, and vehicles out of control all emit sounds that might trigger fear within the listener. But it seems that "[w]ith continued exposure to the noise [...] man [...] will adapt more or less completely, i.e., will cease to show arousal responses" (Kryter 1972: 389). This, of course, only applies to sounds that don't do permanent damage to hearing or have other, non-acoustic effects (see 2.2.2).

Unwanted sound often is a synonym for noise. "[S]ound which is undesired by the recipient" (Ingerslev 1972: 596) "is an aversive component of a setting" (Jones, Chapman, and Auburn 1981: 54), or "disturbs a reasonable person of normal sensibilities" (City of Austin 2008 § 9–2–1 (4)(b)). It "can become an inescapable presence" in the life of the urban dweller (Greater London Authority 2004: 2) and consist of various sources. Aside from qualities like tonality or temporariness, noise is mainly defined by sound level (Ingerslev 1972: 596). For listeners, not being the "operator (s) of the source" of a sound and having no impact on its emission can add to the connotation of a sound as being unwanted (Maris et al. 2007: 3491).

Outdoor noise will always be a problem in urban areas, as there are too many sound sources constantly emit acoustic signals and creating an inharmonic interplay . In contrast to indoor sound, which often can be controlled by the listener, and consists of fewer sources, outdoor noise might be regulated but can never be transformed into a soundscape every listener will appreciate (Ingerslev 1972: 595). In the urban setting, most outdoor sounds are created by traffic.

> The spatial distribution of traffic noise in a city is related to the distributions of traffic volume and urban density, and also to urban form (Salomons and Berghauser Pont 2012: 2).

Highways, with their large number of vehicles driving rather fast, have a particularly high sound level. Ways of reducing traffic noise can be using silent asphalt, noise protection walls, and other constructions (see 4.2), but in many cases protection is still missing. "To the bystander, the city expressway is a threat or at least a nuisance, and from his standpoint avoidance of problems should be the basis of many of the basic sensuous criteria" (Banerjee and Southworth 1990: 563). Industrial sounds also contribute to unwanted soundscape. Construction sites in urban areas not only produce noise through vehicles and heavy machinery, but also by demanding alternative routing and generating traffic jams. Recreational activities are a third major source for urban noise. Sports like basketball, skateboarding, and live music performed in downtown venues create problems for the acoustic environment. "Most of this noise is generated by people traveling to and from work, by the transport of goods from place to place, and by the legitimate use of domestic appliances and leisure equipment" (Institute for Environment and Health 1997: 5). An interesting term regarding the experience of urban unwanted sound is "community annoyance" (Lim et al. 2008: 771), which refers to a common understanding within a community of what should be labeled "unwanted sound". As section 4.2.3 will show, the "exposure to civil aircraft noise" (Lim et al. 2008: 766) can foster shared beliefs and collective action like joint protest movements. People living in the same neighborhood are usually exposed to a similar set of sounds. Alam et al. investigate acoustic discomfort of high-rise dwellers in Singapore resulting from "several community noise sources in close proximity. These include food center, children playground, soccer playground, basketball playground, waste disposal truck, etc." (Alam et al. 2010: 3511). Especially for residents living in high-rise buildings, "community noise was one of the prime sources of noise" (Alam et al. 2010: 3511). As discussed earlier, sound level is only one aspect of a sound. As Alam et al. show, "except for waste disposal truck noise, the noise exposures due to other community noise sources were within the established acceptable limit" (2010: 3518), but they trigger annoyance nonetheless. Overall, there are seven different types of sources for outdoor noise. This list might be modified, subcategorized, or extended but gives a good overview of the cacophony the urban soundscape consists of:

> (1) motor vehicles; (2) aircraft; (3) railways; (4) industrial and commercial enterprises; (5) demolitions, construction, and road works; (6) racing tracks,

> amusement parks, night clubs, discotheques, etc.; (7) local human activities such as noisy persons, slamming of car doors, lawn mowers, and the like (Ingerslev 1972: 597).

Indoors, there are noise sources that add to the urban soundscape, like dishwashers, air-conditioning systems, noise emitted from neighbors, and home entertainment systems playing music, movies, and video games (Ingerslev 1972: 596).

One way to not listen to an unwanted part of a soundscape is to mask it. Whenever a listener is not operating a sound source and can therefore tune it out, covering noise is a popular way to avoid it. "Noise cannot, of course, be totally eliminated, but noise can usually be reduced to an acceptable level" (Ingerslev 1972: 595). Positively connoted sounds are used to soften the impact of unwanted sounds. Nature sounds – which Schafer's studies in acoustic ecology understood as the most pleasing (see Schafer 1994) – enhance the pleasantness of the acoustic environment.

> Adding fountain sound reduced the loudness of road traffic noise only if the latter had low temporal variability. Conversely, adding bird sound significantly enhanced soundscape pleasantness and eventfulness (Coensel, Vanwetswinkel, and Botteldooren 2011: 148).

The reason for this positive impact might be that the overall soundscape consists of proportionally more wanted sounds than before. Another explanation could be "that auditory attention is drawn to that [wanted] sound of the mixture which has the highest temporal variability in sound level, and that the resulting lower amount of attention paid to the other [unwanted] sounds leads to a reduction in perceived loudness" and annoyance (Coensel, Vanwetswinkel, and Botteldooren 2011: 152). Unwanted sounds themselves can also interfere "with the reception of wanted auditory signals" (Kryter 1972: 389). Communications, for example, can be masked by traffic noise. Airplanes flying over a neighborhood, sound emissions from a nearby highway, and the loud beats of local music venues can make conversations – at least temporarily – impossible.

> The degree of interference with communication varies with the amplitude and frequency of background noise, the distance between speaker and listener, the speaker's vocal effort, and the hearing acuity of the listener (Ng 2000: 376).

Many studies have highlighted the negative effect of traffic noise on communication, which in turn has a direct impact on social interaction (Greater London Authority 2004, Institute for Environment and Health 1997, Jones, Chapman, and Auburn 1981, Ng 2000).

There are a number of non-auditory effects that can have an extreme impact on quality of life (Byrne, Michael, and Tufts 2011: 1507, Tassi et al. 2012: 45).

> Non-auditory effects of noise can be defined as 'all those effects on health and well-being which are caused by exposure to noise with the exclusion of effects on the hearing organ and effects which are due to the masking of auditory information (i.e., communication problems)' (Smith 1991: 49).

Working, for example, can be disturbed due to sound (Institute for Environment and Health 1997: 9). A loud environment is deleterious not only for the ears but the quality of performed tasks (Jones, Chapman, and Auburn 1981: 43). Even if a sound in the workspace is perceived in an individual's perception as wanted – like music coming from a radio – performance can go down (Smith and Broadbent 1991: 11). Noise is a "non-specific stressor" that can influence actions and the human body (Smith 1991: 49). Cardiovascular diseases like an "increase in heart rate" can be the result of unwanted sound (Institute for Environment and Health 1997: 10, Stansfeld and Matheson 2003: 243). A higher catecholamine secretion can be registered (Stansfeld and Matheson 2003: 243), thus "hormones, adrenaline and noradrenaline, are raised in working populations exposed to high levels of noise" (Institute for Environment and Health 1997: 11). Psychological symptoms also exist, such as physical reactions like eye-blink responses to sudden noises (Kryter 1972: 389). Tassi et al. investigated the effects of "long term exposure to nocturnal railway noise" and allude to an "increased reaction time" of the affected persons. In some cases, noise is held responsible for preterm birth (Institute for Environment and Health 1997: 11), decreased reading, long-term memory problems (Stansfeld and Matheson 2003: 243), and a less safe workspace due to the increased risk of missing warning signals (Byrne, Michael, and Tufts 2011: 1507). General annoyance is "[t]he most widespread and documented response to noise [...], which may include fear and mild anger" (Institute for Environment and Health 1997: 12). The strongest non-auditory effect of noise might be sleep disturbance (Kluizenaar et al. 2011: 1936, Ng 2000: 376, Smith 1991: 49).

> Noise exposure during sleep may cause awakenings and change in sleep stages. Road traffic noise may also increase the time taken to fall asleep (Institute for Environment and Health 1997: 9).

Some studies explicitly focus on the assumed effects of noise on children – including their sleep patterns – (Öhrström et al. 2006: 116, Stansfeld et al. 2009: 206). Overall, "[t]he evidence for effects of environmental noise on health is strongest for annoyance, sleep and cognitive performance" (Stansfeld and Matheson 2003: 253, see also Institute for Environment and Health 1997: 89).

2.2.2 Pain and Torture

There are extreme ways sound can impact the listener negatively. Being annoyed by traffic noise is one thing, getting tortured in a permanent detainment camp at

Guantánamo Bay something totally different. Between annoyance and torture lies a wide variety of ways sound can be experienced. Aside from its other qualities, sound directly impacts the listener due to its level.

> Exposure to continuous noise of 85–90 dBA, particularly over a lifetime in industrial settings, can lead to a progressive loss of hearing, with an increase in the threshold of hearing sensitivity (Stansfeld and Matheson 2003: 243).

Above a certain decibel level, hearing loss occurs and the experience is very painful. But being deafened by sound levels around 110 to 130 decibels – the average loudness of a jet engine – is not the only result of exposure to such loud sound.

> Steady noise above 110 dB (approx.) can cause some temporary and permanent (after years of exposure) changes in size of visual field and noise above 130 dB can cause mystagmus [sic!] and vertigo (Kryter 1972: 389).

Painful sounds are not always linked to a source. Some people experience sound "or noise in the absence of real acoustic stimulation" (Hobson, Chisholm, and El Refaie 2010: 1). Tinnitus "is not heard by anyone else but the sufferer" (Hobson, Chisholm, and El Refaie 2010: 2). People who experience such sounds have almost no chance of getting cured. Masking – like Muzak's acoustic perfume to tune out the unwanted background sounds of a locality – might be the only help to cover the painful sound.

In the history of mankind, legend, myth, and religion tell about painful sounds and the active use of them to torture others. One of the oldest examples would be the sirens of Greek stories like Homer's *Odyssey* (Johnson and Cloonan 2009: 31). The voices of the sirens caused the death of numerous sailors, as did the chant of the German *Lorelei*, which "provoked suicidal melancholy which drove men to drown themselves in the Rhine" (Johnson and Cloonan 2009: 31). In the Bible, sound occurs both to announce catastrophes and as the cause of destruction. An example of the former can be found in Revelation 8:7:

> The first angel sounded, and there followed hail and fire mingled with blood, and they were cast upon the earth: and the third part of trees was burnt up, and all green grass was burnt up (Revelation 8:7, King James Version).

The collapsing walls of Jericho illustrate destruction:

> [A]nd it came to pass, when the people heard the sound of the trumpet, and the people shouted with a great shout, that the wall fell down flat, so that the people went up into the city, every man straight before him, and they took the city (Joshua 6:20, King James Version).

In the twentieth century, the rise of technology was responsible for a new quality of painful sounds. The two world wars – with their machine guns, grenades, and bombs – brought acoustic terror to the soldiers on the front lines. Hearing the constant sound of war had such an impact on listeners that Johnson and Cloonan talked about a "new regime of noise" (2009: 52). During WWII, Nazi Germany used sound in concentration camps to demoralize and mock detainees. Prisoners had to listen to certain pieces of music containing intimidating lyrics (see Johnson and Cloonan 2009: 71–3). Another example of spreading fear through music is the radio broadcasts used in Chile during the coup against the Allende regime, September 11, 1973. "During the coup, the military temporarily replaced television and radio programming with Nazi march music" (Neustadt 2004: 129). The existing, pro socialist music of the *Nueva Canción* (new song) was forbidden and artists – e.g., *Victor Jara* – were tortured and murdered.

Contemporary uses of sound in the context of pain and torture are numerous. Music often accompanies or inspires murder and massacres. Johnson and Cloonan describe several cases:

> [T]he night before Kimveer Gill rampaged through Dawson College, Montreal in September 2006, killing one student and wounding 19 others, he wrote on his blog that he was drinking whisky and listening to heavy metal. In Australia in 1987, when one Gladys Dickinson asked her son to turn down the Dylan album 'Desire' at 4 a.m., he trampled her to death to the accompaniment of the song 'One more Cup of Coffee for the road' […] (2009: 74).

Another example is a relatively new technology called Mosquito. "The Mosquito is a device that emits a high frequency sound wave designed to cause discomfort to youth, but to be inaudible to adults" (Holmes 2012: n.p.). Through a painful sound that is not audible to adults over about 25 years, teenagers are targeted to leave public spaces where they are not wanted. Locations like playgrounds are equipped with the acoustic pro-active security system to help prevent loitering and vandalism. The exclusive provider of the technology for the USA states on their Web site:

> Make your playground a no-loitering zone with the SonicScreen, featuring exclusive Mosquito technology that discourages disruptive after-hours gatherings by emitting an ultrasonic tone that's extremely unpleasant to those between the ages of 13 and early twenties. The SonicScreen can also be tuned to a lower frequency for ages 19–25, where loitering by older age groups is a problem. [T]he SonicScreen is undetectable by children under 12 or pets (Miracle Recreation Equipment Company).

Mosquito technology does not seem to have the effect "of hearing loss for adults or young people as a result of [the] exposure", though it is still unknown "what [kind of] impact high-frequency noise has on unborn children" (Parliamentary

Assembly 2010: n.p.). That it does have an impact seems to be undoubted. "The device causes harm to pre-verbal babies and small children against which parents are unable to protect" (Holmes 2012: n.p.). If the parents are not able to hear the disturbing sound, they do not why the children suddenly start to shout or even cry. The sound emitted by the Mosquito device is pulsating, "with a sound pressure of 75–95 dB and a frequency range of 16–18.5 kHz" (Parliamentary Assembly 2010: n.p.). Holmes (2012: n.p.) notes that the 'Mosquito' device "contravene[s] the United Nations Convention on the Rights of the Child, in particular regarding health and safety". It also interferes with "Article 8 of the European Convention on Human Rights, which protects the right to respect for one's private life, including the right to respect for physical integrity" (Parliamentary Assembly 2010: n.p.).

The violation of human rights is an ambivalent term when it comes to the war on terror. The United States of America runs detention camps in Guantánamo Bay, Cuba, where extreme interrogation techniques like waterboarding are known to be performed. In addition to a wide variety of torture methods, both physical and psychic, sound is used to affects enemies of the state. "[M]usic can be delivered by and within a socio-political order in a way that functions to humiliate, disturb, disorient and to torture" (Johnson and Cloonan 2009: 147) In contrast to the political uses in Chile and Nazi Germany described above, "music can become itself the direct agent of violence" (Johnson and Cloonan 2009: 148). Lyrics might be the dominant sound quality heard during transfer the procedure to a facility, when new prisoners are blindfolded, dressed in tracksuits and diapers, and music is used to mask the surrounding acoustic environment.

"Earphones would be placed over his ears, through which music would sometimes be played" (International Committee of the Red Cross 2007: 6). The sound level is also used as a form of "[m]usical violence" when the prisoners are kept in detention camps – link Camp Delta and Camp Echo in Guantánamo Bay, Cuba – to "exercise [...] power over someone else and the soundscape" (Johnson and Cloonan 2009: 147). Loud music is often named when it comes to techniques of torture. A quote from a former detainee of an American detention camp in Afghanistan illustrates this point:

> [V]ery loud 'shouting' music was constantly playing on an approximately 15 minute repeat loop 24-hours a day. Sometimes the music stopped and was replaced by a loud hissing or crackling noise (International Committee of the Red Cross 2007: 15).

Torture through sound is often accompanied by other methods:

> I was then placed again in the tall box. While I was inside the box loud music was played again and somebody kept banging repeatedly on the box from the outside (International Committee of the Red Cross 2007: 30).

The amount of time prisoners are exposed to unwanted acoustic irradiation varies. In some cases 16-hour interrogations were conducted using "various types of sound" (Committee on Armed Services 2008: 143). Other examples consisted of "24-hours [of music] each day throughout the three weeks I was held there" (International Committee of the Red Cross 2007: 31), and even "24-hours a day throughout the one year period" (International Committee of the Red Cross 2007: 15).

The absence of sound (see 2.4) can also be used for torture. "[D]eprivation of light and sound in living areas" for several hours affects the brain as the loss of sensory input leads to visual and acoustic hallucinations (Committee on Armed Services 2008: 115). Deprivation of the environmental soundscape can be achieved by tuning out every sound source, usually by putting a person into a soundproof chamber. Another way to take away *normal* acoustic stimuli from the listener is to mask them with white noise.

> The BSCT [Behavioral Science Consultation Team] memo described using fans and generators to create white noise as a form of psychological pressure (Committee on Armed Services 2008: 52).

In detention camps like *GTMO* (Guantánamo), the intention of exposure to loud sound, music, and white noise is to:

> 'disorient him [the detainee] and establish fear of the unknown' and emphasize […] that 'the rules have changed' and nobody knows he is here (Committee on Armed Services 2008: 137).

As documents like the "Inquiry into the Treatment of Detainees in US Custody" (Committee on Armed Services 2008) show, sound is often used to torture detainees. Contrary to visual stimuli like crude light 24-hours a day or the deprivation of it, sound seems to be harder to grasp as a means of torture. There is an official request by a MG (major general) Miller "to use sound modulation in interrogations [blackened]" (Committee on Armed Services 2008: 142). This request surprises the answering committee, because "loud music [is often considered] a form of environmental manipulation which had [already] been authorized by the Secretary of Defense in April. MG Miller's written requests for authority to use sound modulation in the [blackened] interrogations are at odds with that understanding" (Committee on Armed Services 2008: 142), as he seems to be unsure if such use of sound is legal.

Playlists indicate that the most important quality of sound for torturing people is its level, as songs of artists like *Britney Spears* and *Metallica* have little in common. Bands like *Massive Attack* and *R.E.M.* have officially protested the use of their music in interrogation, while other artists are proud to help prevent future terrorist attacks. Whatever tune is used to make a person cooperate, playing back loud music for numerous consecutive hours has the most extreme effect on detainees who – like any other person – cannot just close their ears and tune out unwanted sounds.

2.3 Wanted Sounds

> If we must be distracted 10 or 20 times each day, why not by pleasant sounds? Why could not everyone choose his or her own telephone signal? (Schafer 1994: 242).

While times have changed since Schafer asked for individual ringtones, and the environment has gained a wide variety of polyphonic sounds from handheld devices like tablets and smartphones, there are still sounds that appear to be more pleasant than others. Schafer himself has a clear belief that nature sounds are the most pleasant in the trichotomy of nature, human, and mechanical sounds. In an international sound preference survey, he tested 501 people in four different cities/countries. The most pleasant sounds named were brooks/rivers/waterfalls, birds, and a breeze – all nature sounds. The most unpleasant ones were traffic noise and the sounds of specific vehicles – both mechanical sounds (Schafer 1994: 268–9). A classification of traffic as noise indicates that early acoustic ecology generally prefers nature sounds.

> Nevertheless, no sound in nature has attached itself so affectionately to the human imagination as bird vocalizations. In tests in many countries we have asked listeners to identify the most pleasant sounds of their environment; bird-song appears repeatedly at or near the top of the list (Schafer 1994: 29).

Schafer's trichotomy of individual perception is not the only way to find out about the comparative pleasantness of sounds. Social context and sound qualities like sound level can also be determining factors. And if sound is evaluated as being pleasant the question arises: To whom or for what purpose?

Section 2.3.1 will start with pleasantness (Kang 2007), individual perception (Bruner 1990, Jackson 1968), and the social and cultural context (Fraedrich and King 1998, Kastner 2008). The effects of background music are highlighted (Billings 1990, Chebat, Chebat, and Vaillant 2001, Yang and Kang 2005), and the specific form of Muzak (Lanza 1991, Vanel 2008) is of particular interest. The impact on shopping behavior in open-air markets (Guéguen et al. 2007) and malls and supermarkets is described subsequently. Here, background music may be replaced by loud foreground music, always depending on the target group and range of goods (Milliman 1982, Turley and Milliman 2000, Yalch and Spangenberg 1990). Like stores, restaurants try to satisfy customers through music, not only to shorten individual perception of waiting time, but also to influence the experience of eating (Caldwell and Hibbert 2002, Milliman 1986, Wilson 2003).

Sound is recognized to have the potential to positively influence the architecture of a locality (Gifford and Ng 1982, Imaah 2004).

In section 2.3.2 active listening and the more dominant role of sound is described in relation to rituals and reaching states of ecstasy, rapture, and trance. The rather light positive impact of sound in festivals (Duffy 2000) increases when music is understood to contain a *persona* (Robinson and Hatten 2012: 71). The emotional impact of music rises with its potential to bridge the gap between cultural (Kun 2000) and spiritual borders (Anderson 2002). Musical structure like rhythm and melody may aid in the attempt to reach ecstasy (Finney 1947, Jackson 1968), but there does not seem to be a deterministic relationship between music and trance (Becker 1994, Rouget 1985). Another example of the positive effects of sound can be seen in the Stambeli ritual, where music attracts healing spirits who cure a possessed person (Jankowsky 2007). The most extreme example of the positive impact of sound may occur in the Brazilian Candomblé rituals, where music is the key factor in evoking the Gods (Brazeal 2003).

2.3.1 Nice Sounds

"The appearances of their favourite sounds make people feel more pleasant" (Kang 2007: 72) Thus, sound is used in both urban open public spaces and buildings to create a positive acoustic atmosphere. Wanted sounds are those that do not disturb or even attract listeners. Nice sounds can be those that are heard but not actively paid attention to. In comparison to unwanted sounds, pleasant ones can be louder without being annoying.

> When introducing a pleasant sound in urban open public spaces as a masking sound, there is always a concern regarding its level. [...] [I]t can be seen that this level could be 70 dB(A) or even higher (Kang 2007: 68).

But what are nice sounds?

> [P]leasantness of sounds [...] is based on an instantaneous overall impression emerging from various sound attributes as well as individual preferences and experience (Kang 2007: 5).

The individual approach listening is the first of three important factors for experiencing sound as pleasant, nice, or wanted. There are listeners who prefer free jazz and others who favor heavy metal. Some people love the sound of a Harley Davidson running; others prefer the acoustic insulation of a Mercedes Benz. Emotional meaning is ascribed to certain sounds (Bruner 1990: 99). There may be sounds from our youth or a special music album that trigger certain feelings. Jackson (1968: 297) believes that listeners themselves decide whether music has an influence on them or not. "Physiological variables (auditory fusion, peceptual frequency, noise and modulation), [and] demographic variables (age, gender, eye-color)" are used to explain individual differences in experiencing sound (Fraedrich and King 1998: 131). Studies show that "the majority of people are left-ear dominant" (Fraedrich and King 1998: 136) and that "[c]onsiderable differences have been found between the subjective evaluation of sound level and the acoustic comfort evaluation" (Yang and Kang 2005: 228).

The social and cultural context is a second factor in how sound is experienced. "Both speech and musical sounds are socially patterned and hence can vary from society to society" (Jackson 1968: 298). Learning to speak and understand one's own mother tongue is an example of this phenomenon (see 1.1.1). As Waldenfels points out, all other languages become alien languages when choosing the own mother tongue (1999: 184). Words, grammar, intonation, even meaning vary and predetermine the acoustic experience. Like a music album that has to be listened to several times to get accustomed to it, pleasantness of sound is related to "many different associations such as where it was first heard, with whom it is heard, and what experiences occurred" (Fraedrich and King 1998: 130). The concrete situation often affects the experience. If the sound of an alarm clock indicates a pie being ready it is certainly perceived more positively than if it is acting a wake-up call (Kastner 2008: 81).

Sound qualities, as a third factor, are characteristics such as sound level, "tone, pitch, frequency, [...] and modulation" (Fraedrich and King 1998: 131). There is a wide variety of studies regarding shopping behavior and restaurant atmosphere that investigate these qualities. Marketing and psychology are typical disciplines dealing with sound qualities and the positive impact they can have on the listener, and there are even "experimental economists to examine effects of music in economics decision making" (Fujikawa and Kobayashi 2010: 20). Acoustic comfort is generalized and abstracted from the individual experience. There is a focus on the artificial production of favorable soundscapes and environmental sound sources that positively affect the experience (Yang and Kang 2005: 228).

> Background music has gotten so pervasive and sophisticated that it is no longer limited to syrupy strings [...]. We now have avant-garde 'sound installations' permeating malls and automobile showrooms, quaint piano recitals comforting us as we wait in bank lines, telephone techno-tunes keeping us complacently on hold [...] (Lanza 1991: 42–3).

Environmental music that is supposed to stimulate consumers in a positive way is a phenomenon known since the early 1920s (Vanel 2008: 94). The name of the company that developed the concept has become a synonym for background music: Muzak (Lanza 1991: 43). "Background music is a tool for increasing sales and enhancing positive attitudes toward the store" (Chebat, Chebat, and Vaillant 2001: 115). The effects of background music on shoppers are not easy to detect. There are many variables, like the color of walls and the architecture of a store, employees' behavior, and the price range and collection of offered products, that co-determine the shopping experience.

> [T]he effects evoked by store atmosphere are primarily emotional states that are difficult to verbalize. These emotions are temporary and therefore difficult to recall accurately (Billings 1990: 3).

Within that multi-sensory-input store environment, sounds cannot be too fancy but should rather be simple so as to not confuse the target audience (Fraedrich and King 1998: 136). The wanted sounds should not be too pleasant either because "it may […] hinder their persuasive efforts if the music is considered as not fitting in the sales encounter" (Chebat, Chebat, and Vaillant 2001: 122). The best Muzak is described as being the least disturbing: sounds that are heard but not actively listened to (Lanza 1991: 43). It is supposed to "enhance the atmosphere of" a store or open-air farmers market (Chebat, Chebat, and Vaillant 2001: 115) and make potential customers "stay longer" (Guéguen et al. 2007: 268). Background music is not wanted by companies to sell more goods. Shoppers are supposed to like it not only because of its function as acoustic perfume to cover unwanted sounds – like conversations of other customers – and thus enhance the acoustic environment, but also because of "its evocative power – memories related to music" (Chebat, Chebat, and Vaillant 2001: 121). In the end, background music like Muzak is a permanent sound source that shoppers cannot turn off or tune out. Thus, it might be perceived as unwanted noise, where "ears […] are constantly being sabotaged and indoctrinated" (Lanza 1991: 51). But producing negative attitudes is obviously not the intention of Muzak (Milliman 1982: 86), not in a store nor in a restaurant.

Music and ambient sound are probably the most inquired-about stimulus for customer-relations (Ng 2003: 444). Turley and Milliman analyze 16 different studies that cover the issue (Turley and Milliman 2000). Both methods and findings vary, but there are some shared discoveries about the impact of sound on shopping behavior. One significant effect of music is its relation to the amount of time customers spend in a store (Turley and Milliman 2000: 206). Demographic variables like age lead to assumptions regarding preferred music genres that are specifically used to stimulate a longer stay.

> When shoppers were exposed to music that they normally listen to […], they reported spending less time in the store than they had intended relative to when they listened to music they do not usually select […] (Yalch and Spangenberg 1990: 36–7).

The positive effect – to have shoppers stay longer – can be achieved by using background music that is not too interruptive. But foreground music, even played particularly loud, can have the same impact. Most importantly, the target group's music preferences have to be known to avoid inverted, negative results

(Milliman 1982: 91). Alpert and Alpert examine "the relationship of music and consumers' moods, attitudes, and behaviors" (1989: 485). They demonstrate how happy music "produces the highest average mood" in comparison to no music and sad music (Alpert and Alpert 1989: 489). Milliman varies tempo to differentiate between an artificial soundscape based on no music, slow tempo music, and fast tempo music (Milliman 1982: 87). Overall, there is an "importance of nonmusical sounds as they relate to business and marketing" (Fraedrich and King 1998: 127) that applies to in-store shopping and media business models. Aside from podcasts, radio is the only significant audio medium that uses positive connoted sounds to generate a "more favorable attitude toward the commercial" (Miller and Marks 1992: 83). In addition to a narrator's speech, sound effects can "produce greater mental imagery compared to message-only commercials" (Miller and Marks 1992: 88).

Wanted sounds in media commercials or in-store settings should vary to attract a higher number of customers and match particular goods (Yalch and Spangenberg 1990: 31). In restaurants, sound has to be used according to targeted customers and cuisine. Aside from tempo, which has been found to "affect...patrons' dining speed" (Milliman 1986: 288) genre is the most important factor. Wilson shows that popular music, classical music, and jazz are suitable for modern localities. In contrast to easy listening Muzak, music like jazz "creates the perception of an upbeat or upmarket environment" (Wilson 2003: 107). Consumers spend more time dining when hearing pleasant sounds, which generally results in more money being spent on beverages.

> Understanding of the effects of music is particularly useful to service managers, as this element of the environment is relatively inexpensive and easy to control (Caldwell and Hibbert 2002: 895).

The active use of pleasant sounds may not be as fundamental to a restaurant's success as its food quality or interior design (see Gifford and Ng 1982), but sound is an important element of the atmosphere, going along with the whole architecture of a locality (Imaah 2004).

2.3.2 *Acoustic Ecstasy*

Nice sounds that are supposed to encourage shoppers and customers to buy more goods, feel better, and thus spend more time at a certain locality are the lightest form of sounds that people feel good about. But the wanted sounds of nature and background music are only one part of positive connoted sounds. Others can even lead to ecstasy. Rhythm plays a huge part in the experience of music.

> Rhythm then can be defined as the 'lived' experience of music, where the performer and listener inhabit and give life to music (Duffy 2000: 55).

In contrast to shopping music, which does not have to be in the foreground, music festivals celebrate music and active listening and living. Duffy describes the festival as "a site of intensification" (2000: 52) where rhythm plays a role to "control space and time" (2000: 55). Music is filled with meaning and evokes emotions that listeners can understand and experience.

> We claim that sometimes music can appropriately be heard as containing a 'persona', a fictional or virtual agent whose emotions are expressed in the music (Robinson and Hatten 2012: 71).

While emotions can also be felt while hearing the nice and harmless sounds of supermarkets and malls, active listening to music might trigger "explosive emotional results" (Robinson and Hatten 2012: 104). Audiotopia, as a musical space of difference, has already been mentioned (see Chapter 1.4). The aural border between Mexico and the United States of America is a creative context, where music evokes and transports emotions. Sound becomes the acoustic manifestation of the American north-south divide, an expression of different cultural contexts and ways of life (see Kun 2000). Recorded music is a way to both represent and overcome borders. It is a chance to get somewhere else, not physically but spiritually (Anderson 2002). In 1947 Finney discussed the ability of music to engender ecstasy and rapture. She investigated seventeenth-century society and came to the conclusion that music was believed to have a direct impact on the soul.

> [T]he idea that music could refine the soul to ecstasy was not only a subject which appealed to the metaphysician or the poet. It came to be [...] a basic factor in the serious, even heated, arguments over the use of instrumental music in church (Finney 1947: 186).

Neuronal, chemical, and psychological explanations are used to better understand the assumed link between music and ecstasy. Rhythm plays a major part in it, as constant vibrations of the brain are believed to stimulate ecstatic conditions.

> The [...] question is why [...] [rhythm] should produce psychological effects at all? The answer seems to be that external rhythmic stimulation affects the natural brain rhythms, thus giving rise, in certain cases, to abnormal psychological states (Jackson 1968: 297).

In her inquiry on music and trance, Becker comes to a similar conclusion:

> Music, by stimulating auditory neurons that successively stimulate neurons in the other parts of the brain, helps to establish a particular configuration of activated and unactivated bundles of neurons and activated and inhibited

> neurotransmitters. For some persons, this complex and combined activity results in a state of trance (1994: 50).

The connection between music and trance – defined "as a state of mind characterized by intense focus, the loss of the strong sense of self and access to types of knowledge and experience that are inaccessible in non-trance states" (Becker 1994: 41) – is not deterministic. Listening to music does not automatically lead to trance, while trance can be experienced without listening to music (Becker 1994: 41). But sound can promote trance and events happening while being in trance. Jankowski (2007) illustrates the impact of music in Stambeli, a sub-saharan healing ritual performed in Tunisia. In the ritual, a person reaches a state of trance to get cured by a spirit while music is constantly playing. Music "facilitates the healing process. Its role is to attract the spirits to manifest themselves through induced possession. Once a spirit has taken hold of a host, the music continues to be played for the enjoyment of the spirit [...]" (Jankowsky 2007: 196). Set against the understanding of Jackson (1968), and partly Becker (1994), Rouget does not see music structures like rhythm being of high importance in the creation of trance. But he, like those authors and Jankowski (2007), believes that "[m]usic can undoubtedly sensitize a subject to the call of possession, and can thus contribute to awakening his vocation" (Rouget 1985: 65). Rouget points out "that trance is induced and not conducted" (1985: 322), meaning that music has the potential to act as a trigger for trance more than being an elementary part of it.

Perhaps the peak of the role of music in the creation of ecstasy can be found in the Candomblé rituals to evoke the Caboclos, "Brazilian gods modeled on indigenous peoples and backland cowboys" (Brazeal 2003: 640). Music is not only part of these rituals but a key element (Brazeal 2003: 642). Through music, devotees get possessed by the Caboclos, which express themselves through dance and singing. Music works as a catalyzer to enable the Gods to come into being and talk to their believers. Thus, sound contains mythological power. Whenever the music is altered and new songs are created, the Caboclos can reshape the rituals.

> It is an entire mythology in ballads, quatrains, and couplets. Songs identify the gods and tell their stories. [...] Caboclo worship is characterized by greater innovation and improvisation in music, song composition, dance, and ritual form (Brazeal 2003: 644–5).

The use of sound in rituals like Candomblé may have the most extreme positive effect of sound on the listener. As the section has shown, whether sound is perceived as wanted or not is due to individual perception and nostalgic associations, socio-cultural context, and the various qualities of sound. A high sound level might be accepted, even fostered, because of individual preferences or ritual meaning, but sounds too loud can bother and even hurt the listener. Thus, the absence of sound might be most pleasing for the human ear and its bearer.

2.4 The Absence of Sound

As has been discussed, sounds can have various effects on the listener. Especially in an urban environment with lots of different sound sources, be they natural, human, or technology-made, it is almost not possible to prevent the own body from perceiving constant acoustic input. Thinking of a non-urban context might create the image of a more quiet soundscape, one that Schafer always prefers to city sounds (1994). With technology's help, today a person does not have to flee civilization to find stillness. S/he can put on noise canceling headphones and filter out unwanted sounds.

Especially in an urban environment the absence of sound seems to be desirable, although seldom achievable. The total elimination of sound, total silence, "does not exist in the physical absolute [...]: even when not speaking aloud, man carries on a continuous interior monologue" (Bruneau 1973: 17). But the absence of sound is usually understood as an absence of certain sounds, especially noise, as section 2.4.2 will show in its investigation of the term *tranquility* (Campaign to Protect Rural England 2005, Pheasant et al. 2010, Pheasant et al. 2008). Urban planning and architecture can help to design places within the city where sound is dimmed (Mylott and Dubois Williams 2007). Wanted silence is also part of media, like music (Harris 2005, Kania 2010) and literature (Metzer 2006), and is necessary for any kind of speech or language (Gere 2001). In the Western cultural context, silence is often associated with religion and worship but also appears with negative connotations of power – an example would be silencing someone (Bruneau 1973). Unwanted silence is further described in 2.4.1 (Baranzini and Ramirez 2005, Zembylas and Michaelides 2004).

From a social geographic perspective the question about whether absolute silence exists is not important. The meaning and individual perception of the absence of sound is far more interesting. As the following two sections will show "[s]ilence has positive as well as negative attributes" (Gere 2001: 203). Classification of silence is only possible if silence is thought of as "a concept and process of mind which is imposed by each mind on itself and on the minds of others" (Bruneau 1973: 17). Thus the meaning of silence always depends on its embedment, because "without a good deal of context and implicit understanding, two different kinds of silence can 'sound' the same" (Zembylas and Michaelides 2004: 203). Silence is also a necessary element of all acoustic communication. Language and speech for example only work because of the silent phases in between words (Bruneau 1973: 18).

> We perceive a tune only when silences, however short, mark the beginning and the end; we hear the silence after the last note dies away only because that song has ended and the next has not yet begun. So it is, too, with silence and speech (Acheson 2008: 535).

That silence is never associated with the complete absence of all sound applies to media like film, too: "If you want silence, you have to approximate silence with what's called 'room tone'. It's like quiet white noise. But you can't have zero" (Figgis 2003: 1). Especially in an urban environment, silence never *just is*, but has to be engineered. Holding one's breath, putting on noise canceling earphones, or installing noise insulation to buildings are active way of constructing silence (Acheson 2008: 537). Furthermore, the socio-cultural context co-determines what is experienced as silence and how this silence is connoted (Bruneau 1973: 36).

2.4.1 Unwanted Silence

There are different types of silence that are mostly positive. Unwanted silence is less often described but nevertheless can have extreme effects on the individual. In his paper *Communicative Silences: Forms and Functions*, Bruneau highlights three different appearances of silence: "Psycholinguistic Silence [...]; Interactive Silence; and Sociocultural Silence" (1973: 18). While the first and third forms are generally rated positive, interactive silence contains many awkward moments.

> The embarrassment which often occurs during greetings and partings between acquaintances attempting to establish a more permanent relationship is frequently punctuated by interactive silences (Bruneau 1973: 30).

From childhood on, people are confronted with unwanted silence. Children are asked to be quiet while sitting at the dining table while the parents are talking to each other, or a sibling could wake up because of too much sound. It is used in daily communication to discredit a conversational partner by responding with silence to her/his talk or even refusing a standard greeting. "The silent lack of acknowledgment during a greeting by one or both participants who have been acquaintances or friends can be a strong form of attack" (Bruneau 1973: 30). This type of silence is like a denial of the existence of the conversational partner. Identity that is not only constructed individually but socially needs intersubjective exchange, i.e., some form of communication (Wissmann 2011: 60). As Watzlawick says:

> There could not be a punishment more inhumane than living in the midst of a society unnoticed and unappreciated by all its members (Watzlawick 1995: 21).

Silence is a major tool to produce and demonstrate social power.

> The use of initial silencing strategy by authority or perceived silencing strategy by the subordinate, appears to help assert or reassert interpersonal and group power (Bruneau 1973: 31).

But silence can also be mutual. After a verbal fight two individuals might say nothing to each other for a long period of time. Again, silence is used to indicate a power relation between individuals, e.g., when the one breaking the silence to apologize is the one who has lost the fight.

Milder forms of interactive silence are *nervous silence* and *grave silence* (Bruneau 1973: 32–3). The former indicates uncertain behavior against a special person, social group, or in a specific situation. Whenever someone does not know how to react, silence can express her/his nervousness. The latter might occur after something wrong or inappropriate has been said. Criticizing authorities like teachers, hosts, or employers can immediately result in the silencing of all participants in a conversation.

Unwanted silence can be avoided. Possible awkward moments in a conversation where nobody is speaking can be prevented if somebody fills this silence with small talk. Content-free conversations are often maintained only to avoid moments of silence. "The current educational system in the West is rooted in 'fear of silence', which is one reason the understanding of silence in negative terms prevails" (Zembylas and Michaelides 2004: 208). The paradigm of speaking only if you have something to say does not apply in many day-to-day situations. As many authors report, Eastern culture has a totally different understanding of interactive silence that is more positively connoted.

Another example of unwanted silence occurs in situations where extreme emotions are set free. As "[s]ilence is the language of all strong passions: love, anger, surprise, fear", it can be the only expression possible when "words are difficult to find" (Bruneau 1973: 34). Guilt and shame are also reasons to stay quiet and not say a word. Here, silence is "the manifestation of one's fear of self-exposure through speaking openly" (Zembylas and Michaelides 2004: 203).

While unwanted silence can mean the end of a conversation, it can also occur after a certain statement or a creative expression has been made that does not trigger the assumed response – examples would be speeches, presentations, or any other form of public performance. Silence can also mark the beginning of strong emotional outbreaks. "Deep vengeance, jealousy and resentment can also be very silent-before expressive eruption" (Bruneau 1973: 34). This silence is unwanted because the individual who wants to express certain feelings is not able to do so. It is a holding back, a delay, in the process of communication that can only proceed when silence is overcome.

2.4.2 Wanted Silence

Silence is not necessarily the opposite of communication. Aside from interactive silence, which has been discussed in the context of the unwanted absence of sound, there is also phycholinguistic silence, a vital part of language, especially speech (Bruneau 1973: 23). Silence, in this case, is part of verbal communication. Talking without any pauses is not possible. Phonetic categories (see 2.1.2) can only be deconstructed when there are pauses between them. Silence in between words can also have a special meaning and be put there to express something. The way words are strung together, including pauses, is described as *encoding*.

The speaker actively decides how to present his words. S/he also has to think about how the speech is evolving, which words – including their meaning – have to be used, and where to stop and start a sentence. The listener is receiving the acoustic information and uses the silent sequences to process what is being said. This is *decoding*. Silence also contains meaning per se that is necessary to fully understand speech. "[R]esigning silence to zero-signifier status alone promotes a false understanding of the nature of sign systems" (Acheson 2008: 537). So, "[i]nstead of seeing silence as speech's opposite, we can conceive of it as a part of speech, located on a continuum that puts one in dialogue with the other" (Gere 2001: 206).

In media like literature, silence can be used to communicate certain moods and/or atmospheres. Characters are confronted with silence they might not like. But stepping outside the book and taking the author's position, silence is an always-wanted tool to triggers specific emotions in the reader. Metzer shows that in the works of Samuel Beckett silence is used to extend scenes and raise tension.

> [T]he moments grow more prominent, reaching an unsettling climax for the title characters [...] What disturbs them is a new and persistent 'sound', one that is neither speech nor tone. Silence interrupts their private dialogue, demanding their attention and ours (2006: 332).

To describe silence as disturbing is an interesting fact section 4.5 will investigate further by using an analogy in Waldenfels' phenomenology, where the alien is understood as a disturber of everything that is known and familiar (see Waldenfels 2011).

Like language, music always contains not only sound but silence that:

> refers to places where a notated absence of sound (in later orchestral music, what is known as a 'general pause' or G.P.) creates a rupture in the musical flow (in the full sense of being simultaneously 'abrupt', 'disruptive', and 'interruptive') (Harris 2005: 522).

There is an interesting question about whether silence is supposed to be understood as music or not. Silence before the next chord on a guitar is strummed, the next keys on a piano are played, of a breath is taken before blowing a whistle, horn, or clarinet: Is this part of music? Kania considers "how best to incorporate the concept of silence into a definition of music and consider the possibility of musical works consisting entirely of silence" (2010: 343). If music is defined as the organized form of sound (Schafer 1994) silence would seem to have no part in it. But if we think about examples like the opening of Beethoven's *Symphony No. 5 in C minor* or the pauses in between handclaps in Queen's *We Will Rock You*, silence without doubt is part of these compositions. In 1952, avant-garde composer John Cage premiered his work *4'33"* that only consists of silence (see Kania 2010: 345).

The listener is challenged to listen to music that does not fit into Schafer's former definition of the term. This example describes the maximum of wanted silence in music.

Coming back to the issue of communication, interactive silence doesn't only have negative effects. Oftentimes, the absence of sound is useful to keeping conversations going.

> Interactive silences between persons appear to be highly variable. They are highly bound to the nature of the message sharing process and especially to communicative situations and circumstances (Bruneau 1973: 28).

If the situation is connoted positively, silence can be completely wanted by both speaker and listener. Silent nodding, for example, signals understanding, while a question can also be answered with silence – at least as the immediate reaction – to indicate reflection on the topic. In education silence can be a valuable tool to deepen understanding, to give a child or student the time to think about an idea or lesson. "Silence can [...] be used to help us make sense of thoughts, ideas, emotions, and actions" (Zembylas and Michaelides 2004: 203). This also applies to partnerships, like marriags, where often no words are needed for two people to understand each other. Silence is an indicator of familiarity and intimacy and often goes along with non-verbal communication and perfect knowledge of the other person.

Power relations can consist of wanted silent moments, too. Instead of the forced, unwanted variety, silence "to protect persons in authoritative positions" is a wanted phenomenon (Bruneau 1973: 39). Thereby, silence can also emanate from subordinates:

> Silence is used to show respect [...] For instance, when teachers, clergy, doctors, lawyers, judges, and many others in authoritative positions enter rooms, a characteristic hush falls upon persons in that room. The silence speaks, saying, 'We recognize and acknowledge your position by our silence' (Bruneau 1973: 73).

Related to power is the silence that exists in religion and mysticism (see 4.5). As section 2.3.2 has shown, music and rhythm are part of many religious rituals. Drum-based songs in particular contain various pauses and moments of silence (see Finney 1947, Jackson 1968). In church prayer, silence is a way of communicating with God and is often the only answer to an unknown higher spirit.

> In the Western mystical tradition, silence is an attempt to grapple with the paradox of expressing the inexpressible. Silence is both a sign of awe in the face of the transcendent and a kind of release from the burden of trying to express the inexpressible (Zembylas and Michaelides 2004: 199).

Urban public places are usually not very silent. As too much sound, especially noise, can have extremely negative effects on people (see 2.2), city planning and architecture have to provide basic conditions for at least some locations where it is possible to find some quiet. In public parks and gardens relaxing nature sounds, including water, trees, birds, etc., are part of the soundscape. To create tranquil space in the urban environment "the percentage of natural features present within a scene is [...] a significant influencing factor" (Pheasant et al. 2008: 1456). Parks and gardens are not only attractive because of soothing colors and organic layout but because of the calming effect of the local sounds. "Designed as urban oases, they were places for quiet reflection, spaces of silence where one could avoid urban congestion" (Mylott and Dubois Williams 2007: 1). "Tranquility" is a keyword for many studies on wanted silence in the city. In this context silence never means absolute silence but a minimization of unwanted sounds and an increase of wanted ones. "[A] tranquil area: not necessarily absolute silence but something different from the urban experience, somewhere with an opportunity to hear non-human sounds" (Campaign to Protect Rural England 2005: 41). What a silent or tranquil place should sound like is hard to define. Pheasant et al. suggest that tranquil "environments need to comprise sufficient sensory stimulation to keep us engaged" (2010: 1). But with rising sound levels in the city, silent spots also get acoustically contaminated. Changing recreational activities have also altered the soundscape of silent places, where visitors no longer only quietly walk or sit down to read a book but work out and perform active recreation: "Those seeking quiet can have difficulty finding it among the playgrounds, picnic shelters, and baseball fields" (Mylott and Dubois Williams 2007: 1). As section 3.4.3 will show, the most quiet place in Austin's Zilker Park is actually louder (49.9 dB(A)) than the most quiet one in its Hyde Park neighborhood (42.5 dB(A)) because of nearby streets and highways, numerous joggers, and cyclists. But sound levels around 50 to 55 decibels are still experienced as silent in comparison to the rest of the city.

> It has been shown that in order to achieve a high level of tranquillity [...], the percentage of natural features present should be close to 100 percent and man-made noise sources should be characterized by an LAmax 55 dB (maximum sound pressure level) (Pheasant et al. 2008: 1456).

Even if public parks are created to allow "people to relax, to escape from the stresses and stains of everyday life" (Campaign to Protect Rural England 2005: 3), in high-density agglomerations wanted silence is one of the most precious goods. As Baranzini and Ramirez found out that noise negatively affects rents (2005). Quietness has an implicit economic value; real estate suffers from a significant loss in value when a neighborhood gets contaminated with noise. As section 4.2.3 will show using the example of Frankfurt Airport in Germany, silence is not only a wanted element of urban life but an essential one, and one that must not be neglected.

2.5 The Individual Soundscape

To find quietness and tranquility, people do not have to rely on public places like parks and gardens. In contrast to city planning, every urban dweller can actively influence the soundscape s/he is exposed to by personalizing it. That individually selected music has a positive effect on the listener (DeNora and Belcher 2000) has been described before (Duffy 2000, Robinson and Hatten 2012, Yang and Kang 2005). With the rise of the Sony Walkman in 1979, the commodification of public space and places was taken to a new level. While music in pre-mobile times was enjoyed mostly on the home stereo, Walkman users were able to modify the general experience of the outer environment.

> The auditory and technological nature of Walkman experience transforms our understanding of a range of everyday urban experiences. For example, the auditory look differs significantly from the non-auditory look (Bull 2001b: 191).

The individual selection of songs someone listens to is no longer hearable by others. Music genre as a reference to social status – like jazz music supposedly indicating that the listener is well educated (Ziv, Sagi, and Basserman 2008: 471) – is no longer traceable. Music playing in over-ear, on-ear, or in-ear headphones is not part of a common public environment. And what the Walkman started, the Apple iPod brought to an end: changing the urban soundscape from a mostly public space to a private and fragmented one. In 30 years, about 200 million Walkman have been sold. In its first 11 years, the iPod almost doubled this number (over 350 million).

> The Apple iPod is the first cultural icon of the twenty-first century, representing a sublime marriage between mobility, aesthetics and functionality, of sound and touch-enabling users to possess their auditory world in the palm of their hand (Bull 2007: 1).

As for mobile listening, the car is often described as the "most intensive radio listening experience, perhaps even your most intensive media experience altogether" (Bull 2001a: 185). But in contrast to listening to music in your own car – a space that, without question, is a private one – Walkmen and iPods invade originary public space. The following sections will investigate whether the urban environment is enhanced by this individual commodification (2.5.1) or neglected due to the masking of all environmental sounds (2.5.2). The iPod stands for a post-modern tolerance of the other, while at the same time describing a trend towards increasing isolation and lack of social cooperation.

> iPod culture embodies a dialectical relationship between the desire for an ever-present intimate or personal connectivity and the impoverishment of the social and geographical environment within which it occurs (Bull 2007: 9).

2.5.1 Personalized Soundscapes

The individual soundscape was there long before it went mobile, as Bull points out with his reference to early-twentieth-century German philosopher Siegfried Kracauer (Bull 2001b: 183). Headphones used at home provide the listener with the same kind of private listening experience as the ones connected to a Walkman or iPod. But it may be the contrast of the privateness of earphone listening with the public environment that really intensifies the individual element of the act. Listening to you own music in your own home is less of a statement than doing so in the open where all other people are actively excluded from the listening experience. Headphones indicate the freedom of choice over the soundscape and proclaim the autonomy of the listener (Hosokawa 1984: 166). But:

> [a]utonomy is not always synonymous with isolation, individualisation, separation from reality [...]. Walkman users are not necessarily detached [...] from the environment, closing their ears, but are unified in the autonomous and singular moment [...] with the real (Hosokawa 1984: 170).

In a positive way, music modifies the environment for the listener (Bull 2001b: 181) and enhances the urban experience (Bull 2000: 91, 2001a: 188). As mentioned above, private mobile listening does not have to be restricted to earphones, as the example of "placing [...] a radio within the automobile" shows (Bull 2001a: 188).

> The car is one of the most powerful listening environments today, as one of the few places where you can listen to whatever you like, as loud as you like, without being concerned about disturbing others [...] The car is the 'bathroom' of our time (Bull 2001a: 187).

Despite safety concerns related to listening to loud music in a car – e.g., masking possible signals like the siren from an approaching ambulance – there are several upsides. It makes the daily commute more interesting and helps distract listeners from their work and "daily routine" (Bull 2000: 190). Radio energizes the space of the car (Bull 2001a: 192) and "gives users a sense of companionship. They 'know' their music and never fell 'alone' whilst listening to it" (Bull 2000: 189). In a way, the car radio enriches the landscape the listener is driving through, giving the journey a certain meaning (Bull 2001a: 194). As with all forms of mobile listening in public spaces, in a car "[u]sers often pick music to 'suit' the environment [...]" (Bull 2000: 188). An automobile user describes the situation:

> I wind the window down so I can hear what's going on and sometimes as the sun's setting and I'm in town and I think. Wow. What a beautiful city that I'm living in, but it's always at the same time when that certain track comes on. It's a boost (Bull 2001a: 185).

Adding an individual soundscape to the experience of the urban environment can lead to an almost filmic experience. The listener moves through the urban setting, the lead actor in her/his own blockbuster (Bull 2007: 45). It does not have to be actual film music to evoke the effect of such an experience. Sounds from or associated with TV shows, music videos, and books construct a certain image of the city that is highly individualized (for a discussion on the medial creation of the city, see Chapter 5). One respondent in Bull's study states that the image is:

> her own creation taken from a stock of memories of heroines from films half forgotten, scenes barely remembered or scenes from the multitude of pop videos watched distractedly, or maybe even a trace from childhood novels of romance read from under the bedclothes deep into the night (Bull 2000: 89).

One piece of music works better to create one's own narrative than another piece. The given temporal and cultural setting of the listener is of importance. The selection of individually consumed media reflects a certain context, like the political situation – think about the music of the 1960s and movies like *Easy Rider* – fashion trends, and contemporary literature (see Tagg 1994: 55).

With the rise of the iPod at the beginning of the twenty-first century, mobile listening became almost ubiquitous. After the initial product phase, where white earphones were perceived as a curiosity, dropping prices, a brought product line-up, and higher sales figures have enabled people to acoustically construct urban space the way they want to (Bull 2007: 5). "Users attend to an experience of the city they themselves, in part, construct through their soundtrack of choice" (Simun 2009: 931). With new ways to distribute music over the Internet, desired songs can be downloaded instantly with smartphones, tablets, and music players that come with a permanent connection to online stores. Through individual control of the soundscape, the listener changes the urban environment. "The meaning of city spaces itself derives from the playlists of users" (Bull 2007: 37). By changing the urban soundscape, the listener modifies not only her/his own environment, but also her/his self. The personal stereo lets users feel "a greater sense of purpose" (Bull 2000: 190). But music can only serve as an activator that positively affects mood and behavior if the *right* music is coming from earphones.

> Failure to have the correct music invariably makes the personal stereo dysfunctional, leading users to switch their machines off as a preferable alternative to listening to 'incorrect music'. Incorrect is defined in terms of the sounds not matching either the mood of the user, or their surroundings (Bull 2001b: 185–6).

With the rise of mp3 players that hold thousands of songs, having the right music at your fingertips is no longer an issue. If the collection is too large, organization and ease of navigation can be achieved by putting up a proper playlist – often generated with a single command. “MP3s make available such vast music libraries to users that missing the ‘right’ music is a challenge” (Simun 2009: 922).

Portable music players like Walkman and iPod enable listeners to experience the city in a whole new way. For them, unwanted environmental sounds like traffic noise no longer exist. This opens up the possibility of connecting with the city individually. Through music that combines visual impressions with personal emotions and bodily movement, the listener aestheticizes and theatricalizes the city (Hosokawa 1984: 176).

> This aestheticization creates the world as an imaginary space, a projection of the desire of the user formulated within the cultural remit of the stock of their imagination which is mediated through the attendant sounds listened to (Bull 2000: 188).

To enrich the personal urban experience the choice of sounds is arbitrary. From music to podcasts to audiobooks: every input is possible and fits as well or poorly into the multi-sensory scene as any other piece of audio. Listening while walking has become an urban strategy (Hosokawa 1984: 166), where users do not necessarily have to refuse communication with their surroundings, but sometimes do.

2.5.2 Fading out the Acoustic Environment

The Walkman has been:

> positioned as, on the one hand, the purveyor of choice and increased freedom for the individual and, on the other, as the destroyer of public life and community values. It was this latter representation […] that was the main focus of public debate (Du Gay 1997: 113).

This quote describes the situation in the 1970s, when Sony launched the first Walkman and mobile listening became a pop-culture phenomenon. Never before had so many people been spotted wearing huge earphones, moving to a secret beat only they could hear.

> [T]he city of the iPod user is made up of parallel and privatised soundscapes in which each iPod user looks around the subway compartment at others equally enclosed in their private soundscape, each ‘inhabiting their own reality’ (Bull 2007: 29).

Bringing one’s own soundscape into the public space, as stated above, intensifies the act of private listening. Privateness becomes especially apparent when putting on the earphones in a noisy environment and pressing the play button. The experience changes automatically. "The Walkman [– respectively the iPod –] is doubtless the ultimate object for private listening" (Hosokawa 1984: 168). Public space is transformed into private space, "allowing private 'domestic' pleasures" in the open (Du Gay 1997: 113). The experience of the other is avoided by denying its (acoustic) existence. Space is *cleared* from all unwanted sounds and made easily accessible by the individual through most familiar tunes (see Bull 2000: 189).

> [I]n reconfiguring their relationship to the environment, they [the listeners] also reconfigure the environment itself, negating the very public nature of the urban spaces they navigate (Simun 2009: 938).

The definition of space being *public* is indeed questionable when sound is considered to be a part of it. The individualization of the soundscape affects the perception of space, which no longer contains public environmental sounds – at least those sounds are no longer hearable while wearing earphones and listening to loud music. Tagg notes that with the Walkman, "it is about time adults gained at least some intellectual control over the sounds, both in and outside music, of our society" (1994: 57), but the question about the character of such a society should be discussed at the same time. If the iPod is an icon for the individualization of both the listener and her/his soundscape, it also represents an increasingly social *chilliness*. It seems like users of mobile listening devices live – at least acoustically – in private space, where only individual observation takes place, instead of being in communication with other people.

> iPod culture represents an antithesis to this visualised understanding of cosmopolitan space. If cosmopolitanism remains as an experiential category in urban life, it has migrated from the street to the diverse and often multicultural nature of the user's playlists (Bull 2007: 26).

The city is getting more and more segregated into different genres of different musicians, into different playlists of different listeners. What sounds like a wide variety of choice and the opportunity to express all individual preferences does not result in a cultural cacophony. Each listener only listens to her/his own music, tuning out not only environmental sounds but the sounds of other Walkman and iPod users, too.

> Urban experience undergoes a systematic 'ghettoisation' of which iPod use is the most recent and pervasive manifestation, furthering the technological intervention of subjects intent on urban retreat, transforming the polyrhythmic nature of urban space into the interiorised mono rhythmic sounds of users (Bull 2007: 2008).

In a phenomenological sense, the *own* takes over the *alien* space of daily life. An observer might describe the distance of the listener to the environment as a turning away from *reality*. At the same time the commodification of the urban soundscape is a turning toward the city, which in turn becomes more intimate and familiar (see Bull 2001b: 188, Hosokawa 1984: 171). When Simun asked users of mobile listening devices about their reasons for putting on earphones, he found that:

> 'escaping' is the most prominent MP3 practice emerging from respondents' descriptions [...] MP3 enables users to escape the cognitive, emotional and even physical experience of the environment, as well as unwanted interaction with others (2009: 931).

This statement does not have to be understood in contrast to the enhancing effect a personal stereo can have on the environment. If listeners can "partially pretend that they are not really there" (Bull 2000: 186) because they are listening to a private playlist, this negation of certain elements of the environment can also be seen as an individual upgrade of one's personal surroundings. In the first days of the Walkman, "its presence in the public domain [might have] offended people's ideas about what sorts of activities belonged where" (Du Gay 1997: 155), but fading out the acoustic environment today is far more accepted and relatively *normal*. "This listener seems to cut the auditory contact with the outer world where he really lives: seeking the perfection of his 'individual' zone of listening" (Hosokawa 1984: 167). This zone of comfort cannot be disturbed by unwanted sounds. Thus, "iPod culture is a filtering culture" (Bull 2007: 22). The listener is aware of her/his environment and – while s/he cannot influence the visual experience by restructuring buildings or the city grid – modifies perception with additional sound. This gives her/him, at least partly, control of the environment, as the personal stereo is a "defensive strategy" as well (Bull 2007: 22). As mentioned before, music is a good way to exclude oneself from the world, just like it is a good way to exclude the world from oneself (Simun 2009: 930). The sound quality and comfort of ear- or headphones has reached a level where total acoustic isolation is possible.

> Users might be aiming to block out any external sound that they might otherwise hear in the street or elsewhere. The intrusion of any collection of aural sounds located within space is unwanted (Bull 2000: 186).

By fading out environmental sounds the meaning of the city is decontextualized or – because the personal soundscape always has its own significance – recontextualized (Hosokawa 1984: 171).

> The iPod appears to privilege private life, enhancing the conception of the consumer as an isolated subject. In doing so users appear able, and often desire, to liberate themselves from the perceived oppressive rhythms of daily life (Bull 2007: 7).

Describing listeners as consumers makes the need for a modified soundscape even more comprehensible. It is not only the sounds of cars and construction areas, other people walking and talking, and wind blowing and birds singing that might disturb one's sound experience in an urban environment.

> In a city filled with media messages – from tube announcements […], to the TV adverts on the bus […], to […] 'music blaring from the shops' – the MP3 enables users to, in some ways, 'silently sabotage' the domination of mass culture (Simun 2009: 938).

Especially in cities, with their high population densities, multi-source sound inputs, and acoustic stimulus satiation, audio becomes more important to preserve a certain amount of privacy (see Bull 2007: 32). But what privatized soundscapes do to an urban society that defines itself at least partially by the "shared experience of public space" (Simun 2009: 940) is hard to predict. The disappearance of public space and its replacement by multiple private versions of it result in both individual fulfillment and social coldness.

Chapter 3
Sound in the City

3.1 Soundscape Studies in Acoustic Ecology

"Unless we listen with attention, there is a danger that some of the more delicate and quiet sounds may pass unnoticed" (Westerkamp 1974: n.p.). This quote by Hildegard Westerkamp, member of the *World Soundscape Project*, has a double meaning. It refers to the sense of hearing and the need to sensitize the ears to environmental sounds. It also testifies to the situation scientific research was in, in the 1970s: Sound did not play a major role in it. Truax, also involved in the Canadian World Soundscape Project, describes the situation at that time as:

> an unfortunate lacuna as knowledge about how sound acquires meaning and functions to create social and psychological order [...] would be a significant contribution to a theory on which to base environmental music (Truax 1996: 51).

The study of environmental music – the use of nature sounds to enrich musical pieces or compose entirely new ones – involves an understanding of the importance of sound for all people who are capable of hearing. "Anything in our world that moves vibrates air", says Murray Schafer (1969: 5), widening the research area of sound to include almost any environmental event. As maybe the most important character in soundscape studies, Schafer mainly focuses on nature sounds to sensitize both public and science to sound. The significance of sound can be explained not only because of its various effects on the urban dweller (see Chapter 2), but also from a historical point of view:

> Big noises like cannons, church bells, steam engines and jets have changed history as much as bold proclamations. So have small sounds, pronounced in whispers at clandestine meetings (Bull and Back 2003: 26).

As talk has changed the course of history, language is always dominated by visual analogies (Truax 1978: n.p.). The sound of a word itself can be of importance for its meaning, or even – in the case of an onomatopoeia or a "sound word" found in cartoons or comics – representing this meaning through pronunciation and accentuation (Bull and Back 2003: 30). On the one hand, sound is not easy to avoid. "The sense of hearing cannot be closed off at will. There are no earlids" (Schafer 1994: 11). On the other hand, many sounds are overheard or "attract little attention in the modern world" (Bull and Back 2003: 38), especially if they derive from a non-technological natural environment. If sensitized, hearing becomes listening and even unimpressive sounds are perceived and paid attention to (see Rodaway 1994: 3).

The principal directive of the World Soundscape Project, "which was formed at Simon Fraser University in British Columbia in the early 1970s" (Waterman 2000: 112, see also McCartney 2000: n.p.), is to document environmental sounds "and to promote increased public awareness" (Truax 1996: 54). At Simon Fraser University, Schafer, Truax, Westerkamp, and others recorded sounds and rearranged them into environmental music collages. In this "excellent climate for thinking about and working with environmental sound in context" (McCartney 2000: n.p.) a new research field emerged, today described as *acoustic ecology*. In addition to collecting environmental sounds for composing purposes, different effects of sound on human beings (see Chapter 2) were investigated (Waterman 2000: 112). At Simon Fraser University the study began to create a common sense and both a theoretical and empirical framework, which had been missing to that point. "Thus, when we look to the body of Western research that deals with environmental sound, we find the glaring absence of a coherent knowledge base" (Truax 1996: 51). For this purpose Murray Schafer wrote his book *The Soundscape: Our Sonic Environment and the Tuning of the World* (1994). Looking at recent bibliographies and citations since then, it is clear that book has had the most impact on all follow-up studies in acoustic ecology and neighboring disciplines. Truax describes Schafer's work as "a largely descriptive basis for soundscape studies" (1996: 53) that is still up to date in many respects (see sections 3.2 and 3.3).

One assumption of acoustic ecology is that it is "easy to take the acoustic environment for granted in daily life" (Peterson 1996: 400). Another one is that "environmental sound acquires its meaning both in terms of its own properties and in terms of its relation to context" (Truax 1996: 52). The last statement indicates the dependence of single sound events on the setting in which they emerge (see section 3.1.2), but almost hides the individual element that is necessary for any meaning. "Environmental sound is decoded by the listener" (Truax 1996: 52). All environmental sounds combined are described as the soundscape, which Truax defines by its communicative quality: "I used the term 'soundscape' […] not just as a synonym for 'acoustic environment', but as a basic term of acoustic communication. It refers to how the individual and society as a whole understand the acoustic environment through listening" (Truax 2001: xviii). A more open description comes from Schafer, who explicitly includes media environments in addition to natural ones:

> The soundscape is any acoustic field of study. We may speak of a musical composition as a soundscape, or a radio program as a soundscape or an acoustic environment as a soundscape. We can isolate an acoustic environment as a field of study just as we can study the characteristics of a given landscape (Schafer 1994: 7).

In his understanding a soundscape is everything that can be heard, which makes the term rather difficult to describe. In comparison to its visual equivalent, *landscape*, a soundscape cannot be displayed in a way that captures an entire scenery, as in a photograph. "The microphone […] samples details. It gives the close-up but nothing corresponding to aerial photography" (Schafer 1994: 7).

The sounds of the world are thought of more as a "musical composition" than a noisy cacophony (Schafer 1994: 5). This leads to an artistic subfield in soundscape research: soundscape composition. Environmental sounds are recorded for electronic music, a totally new kind of music: "Little by little throughout the twentieth century all the conventional definitions of music have been exploded by the abundant activities of musicians themselves" (Schafer 1969: 2). The result is a constructed soundscape like the one Schafer produced for the city of Vancouver, Canada (Schafer 1994: 59). Typical sounds are recorded and afterwards mixed and arranged to create a hearable impression of the city. This, in a way, answers Schafer's own question about which sounds should be "preserve[d], encourage[d], multipl[ied]" (Schafer 1994: 4), because recordings like the Vancouver soundscape not simply represent the city of Vancouver in general, but the city of Vancouver in the 1970s. The statement that all sound is music (Schafer 1969: 2), of course, isn't followed by the whole music industry. Canadian soundscape compositions mainly stay in the country. International radio stations – that use music to transport images and imaginations (see Bull and Back 2003: 36) – generally do not spend airtime on soundscape compositions.

The theory that all sound is music implies that the listener is able to deconstruct the soundscape and set single sounds into perspective to form a more or less harmonic sound collage. This individual perspective on sound can be described as "acoustic communication" (Truax 2001: xvii). Thus acoustic ecology is interested in the "relationship between man and the sounds of his environment" (Schafer 1994: 3). In contrast, Hildegard Westerkamp defines music as "environmental when it accompanies activities of daily life. In other words, the activity is the focal point, not the music" (Westerkamp 1988: 1).

Overall, there are too many sounds to ignore. Acoustic ecology's fundamental idea is to learn how to deconstruct the soundscape to make sense of it and let sounds appear less painful to the listener because of their ascribed meanings (Westerkamp 1974: n.p.). Therefore, "we need to develop a new means for describing the sounds we perceive" (Schafer 1969: 30). Schafer answers his own call with the deconstruction and classification of the soundscape's single sounds (see sections 3.1.1. and 3.1.2). Because places are thought of as containing a certain unique soundscape and "sound picture" (Velasco 2000: 23), Schafer's empirical research begins at specific locations that have special meaning for people. Knowledge of place-based sounds is believed to tell us more about ourselves and our relation to our environment: "We reach out to the sounds because we want and need them for orientation and information, for locating ourselves within a place" (Westerkamp 1988: 15).

3.1.1 Soundscape, Composition, Sound Events

At the place where a sound emerges the physical surroundings have to be investigated. Every building, open space, even moving objects have an impact on sound.

> The serious use of environmental sound, then, means to attend to the context and the integrity of sounds, to be aware of the relationships between sounds and their contexts, and to work with a listener's associations and memories of sound environments (McCartney 2000: n.p.).

Association with sound is especially important when the recorded audio material is used to compose a soundscape or a piece of music out of it. Relating to soundscape composition technology has evolved since the beginning of soundscape studies. "No audio technology, no soundscape" (Iges 2000: n.p.) is a simple but true statement that highlights the dependence of soundscape studies on technological devices. Without microphones and audio recorders, there would be no soundscape composition possible. When Peterson described the need for technology, recording was just moving from analog to digital. With the rise of digital recordings soundscape artist got "versatile tools that make it easy to record and sample vast libraries of natural sounds. A composer can then select, modify, and assemble these sound bites into musical compositions" (1996: 401). The benefits of digital recording change not only the quality of recordings but recorded sound itself. White noise and other side effects of analog technology vanish and alter composing possibilities for the artist.

> Truax concentrates on the importance to the composer of the experiences, awareness and perceptions of listeners, and their relationships to the sound environment. These become an integral part of the compositional strategy (McCartney 2000: n.p.).

The vast possibilities of digital recording and post-production have not made the use of "real" sounds obsolete. Soundscape composition "has evolved rapidly" (Truax 2002: 17), but "[a]t present, soundscape composers rely on high quality recordings of environmental sound as source material, since no synthesis methods have been devised which can produce realistic environmental sounds [...]" (Truax 2002: 17).

The use of only "real" environmental sounds suggests that acoustic ecology's soundscape composers could express the natural soundscape in their artistic collages. "But no recording is an exact reproduction of living sound" (Schafer 1969: 45). This is no wonder, especially because constructed soundscapes are not designed to be an exact replication of nature but often "create an imaginary world with processed sounds of various origins [...]" (Truax 2002: 7). Listening to such a composition reveals the unrealistic character created when certain unnatural rhythms occur or a multitude of significant sound signals are strung together. But single sounds can still be recognized. Summarized, soundscape composition is one of the possibilities of working with environmental sound relating to individual perception and cognition of the world:

> The principles of the soundscape composition are: (a) listener recognizability of the source material [...] (b) the listener's knowledge of the environmental and psychological context of the soundscape material is invoked and encouraged [...] (c) the composer's knowledge of the environmental and psychological context of the soundscape material is allowed to influence the shape of the composition [...], and ultimately the composition is inseparable from some or all of those aspects of reality; and ideally, (d) the work enhances our understanding of the world, [...] (Truax 1996: 63).

To make use of environmental sounds for composition the recording has to capture distinctive sounds like the bark of a dog or the ringing of a church bell. The members of the World Soundscape Project differentiate between environmental sound in its natural context and stored recording.

> [A]ny natural sound, no matter how tiny, can be blown up and shot around the world, or packaged on tape or record for the generations of the future. We have split the sound from the makers of the sound. This dissociation I call schizophonia [...] (Schafer 1969: 43).

A recorded sound is out of context but by referencing its surrounding environmental sounds it can establish a link back to its origin (see Schafer 1969: 44, Truax 1996: 50–51).

The natural as well as the composed soundscape feature numerous sounds that are mixed, linked, and interwoven in various ways. If single sounds can be identified easily by the listener, and background sounds do not disturb single distinctive acoustic signals, the environment is described as hi-fi. "A hi-fi environment is one of acoustic clarity, with a favourable signal-to-noise ratio" (Westerkamp 1988: 6). Sound can be experienced clearly and is not clouded or masked by diffuse noise (see Schafer 1994: 272). Generally, urban sound is not classified as being hi-fi. A much higher quality is ascribed to nature or wilderness (Schafer 1994: 43, Westerkamp 1988: 7).

> The quiet ambiance of the hi-fi soundscape allows the listener to hear farther into the distance just as the countryside exercises long-range viewing. The city abbreviates this facility for distant hearing [...] (Schafer 1994: 43).

In contrast to a hi-fi environment sounds can be "overcrowded, resulting in [...] lack of clarity" (Schafer 1994: 272). The lo-fi environment is often linked to electroacoustic sounds of the city. Traffic noise, electronic devices, and loudspeakers add to a cacophony that early soundscape researchers evaluated negatively. "In a lo-fi soundscape individual acoustic signals are obscured in an overdense population of sounds" (Schafer 1994: 43).

To be able to classify sounds of the soundscape and evaluate their individual impact on the listener and the environment, single sounds have to be revealed and correlated to their sources and physical origins. "Single sounds will be studied more attentively with attention paid to the formants of their overtone spectra and to their onset and decay characteristics" (Schafer 1969: 3). Truax defines a single sound or *sound event* as follows: "A sound or sound sequence in its spatial and temporal context as part of a SOUNDSCAPE" (Truax 1978: n.p.). *Sound events* belong to the natural environment while *sound objects* represent the recorded form of a single sound. "[T]he latter is an abstract acoustical object for study [...]" (Schafer 1994: 274), it is not linked any more to the natural environment with all its connotations and meaning (see Truax 1978: n.p.). Schafer describes the sound object as having a pure acoustic existence, without the possibility for the listener to establish links or interpret it (Schafer 1994: 53). Single sound events have this potential to connect the listener to the environment. "They may be singled out for attention if the need should arise, but normally they aren't specifically noticed" (Truax 2001: 25). This means that the definition of what a single sound event is, is basically up to the listener. As a sound always has a source and is located somewhere, the surrounding sounds might drown it out, or – even more important – the listener might not consider the sound meaningful enough to perceive it as a single sound event. The social context codetermines whether a single sound gets singled out or not. Known sounds are more likely to be experienced as sound events as they carry meaning and can act as a symbol – like church bells for Christianity or a siren as a warning sign.

Although we have implied that every sound event in the acoustic community is unique in the information it conveys, it is also true that such occurrences are not random. Over countless repetitions, the images created in people's minds by such sounds and their contexts build up coherent patterns that may be called sound symbolisms (Truax 2001: 80).

3.1.2 Keynote, Signal, Soundmark

> What the soundscape analyst must do first is to discover the significant features of the soundscape, those sounds which are important either because of their individuality, their numerousness or their domination (Schafer 1994: 9).

Before dealing with the deconstruction of the soundscape it is important to realize that sounds are not only generated by the surrounding environment, but the listener her/himself. One's own heartbeat, footsteps on the ground, and rustling of clothing add to the experience of the soundscape. "Initially we become aware of the fact that we are soundmakers simply by moving through the soundscape" (Westerkamp 1988: 15). As stated above, what the listeners hears is determined by individual and social factors. But whatever meaning is ascribed to certain sounds, they can be singled out or at least categorized into three different elements: *keynote*, *signal*, and *soundmark*.

1. The Keynote: "Keynote is a musical term; it is the note that identifies the key or tonality of a particular composition" (Schafer 1994: 9). In soundscape studies the keynote has a different meaning. Keynote sounds are "those which are heard by a particular society continuously or frequently enough to form a background against which other sounds are perceived" (cited equally in Schafer 1994: 273, and Truax 1978: n.p.). Examples would be sounds that are not too dominant to be singled out. In the wilderness one might think of the waves gently reaching the shore, wind slightly moving the leaves of trees, or birds singing their songs in numerous voices. In a city the sounds of traffic would be an example. Within the sounds of rush hour no single vehicle can be isolated. Industrial sounds and electrical hums also count for an urban keynote sound (Truax 1978: n.p., Westerkamp 1988: 14). But the keynote contains softer sounds, too. "A subtle keynote is offered by the sounds of light. Between the soft sniffing of the candle and the stationary hum of electricity a whole chapter in human social history could be written [...]" (Schafer 1994: 59). The keynote can also be linked to the natural environment (Truax 2001: 25). Climate, for example, changes the basic appearance of a soundscape. In winter, snow absorbs part of the sound volume, and the cracking of frozen water also belongs to the category of keynote sounds (see Schafer 1994: 20). Overall, keynote sounds are nothing special or unique but occur exactly when and how the listener expects them to. "Perhaps the most common reason for sounds to be heard in background listening is that they are a usual occurrence, and therefore expected and predictable" (Truax 2001: 25). They, in a way, "underpin[...] other more fugitive or novel sound events" (Schafer 1994: 48). In terms of acoustics, keynote sounds are more of a low-frequency base than high-level pitch. They "stay in the background of perception, [...] [and] do not interfere too greatly with foreground signals" (Truax 2001: 139).

   Keynote sounds contain a social element that makes difficult especially empirical research. Defining sounds as part of the keynote is always based on social connotation and individual evaluation. "A keynote sound is not necessarily a specific type of sound. The term refers more to how it is perceived, that is, in the background of one's perception" (Westerkamp 1988: 14). That means that any sound perceived by one person as part of the keynote might stand out as unique from another. The siren of an ambulance can evoke distinctive feelings or be experienced as an ordinary element of traffic (see Truax 2001: 25). But sound can have an influence on people's behavior (Schafer 1994: 48). In contrast to the constant sounds hearable in the background, the keynote makes its most impression whenever it vanishes. Often it is only then that it is consciously perceived as an indispensable part of the soundscape. Generally, keynote sounds are heard but not always actively listened to (Schafer 1994: 9). Often they "are not consciously perceived, but they act as conditioning agents in the perception of other sound signals" (Schafer 1994: 273). Those other sound signals have different meanings for the listener. Of these special variations the most basic form is:

2. The Signal: "A signal is a sound with a specific meaning, and it often stimulates a direct response [...]" (Schafer 1994: 169). In contrast to the keynote, signals are noticeable sounds that are consciously perceived. Church bells have already been named as examples, as has the siren of an ambulance. Corresponding to the keynote, signals are not signals by their essence but contain both individual and social meaning. If signals are distinctive elements that cannot be overheard, they do not have to be listened to automatically. Bull and Back point out the preference of "busy people" to listen to "signals of activity", for example (2003: 38). As signals give us information about the listener who identifies them as such, they also contain information about their source, as they emanate from loudspeakers, motors, or radios. As part of a certain soundscape they refer to the surrounding keynote. Their reflection on walls and buildings points to the environment, where the signals occur (see Truax 2001: 47). What all signals have in common is their importance for the community. Thus, they are often described as "community signals" (Schafer 1994: 178). Signals refer to the character of a community with the special sounds they provide. It is not surprising to hear cars honking in the downtown area, but not to hear a bike bell on a bike lane. One would expect to hear children crying at a playground and listen to a dog bark at an animal shelter. Depending on buildings, traffic, and social factors like age, certain sounds are representative of a community. Signals also mark the boundaries of communities. Some signals can only be heard in specific areas that are acoustically delimited by the audibility of their signals (Truax 2001: 67). In any case, "[s]ignals are foreground sounds and they are listened to consciously" (Schafer 1994: 10). Sirens from police cars, ambulance, and fire trucks illustrate that signals are "meant to be listened to" (Truax 1978: n.p.). Using an example from visual perception Schafer compares signal and keynote to *figure* and *ground* (Schafer 1994: 275). To study the various forms of signals Truax suggests focusing on five different aspects: acoustic characteristics, subjective perception, historical importance, comparability to keynote and other signals, and symbolic power (1978: n.p.). One special type of signal is the symbol. It has an even richer connotation and meaning and might trigger stronger emotions in the listener (Schafer 1994: 169). Signals of historic importance are singled out as a third soundscape element:
3. The Soundmark: "A term derived from 'landmark' used in soundscape studies to refer to a community sound which is unique, or possesses

qualities which make it specially regarded or noticed by the people in that community. Soundmarks, therefore, are of cultural and historical significance and merit preservation and protection" (Truax 1978: n.p.). The unique character of the soundmark is stronger than that of the signal. While signals generally occur in a higher number, soundmarks are unique. A good example of a soundmark are the bells of Big Ben in London. The inimitable tone sequence of the bells are an acoustic symbol of the capital of Great Britain, just as the Tower Bridge is a visual one. Every community has its soundmarks (Schafer 1994: 239), even if it might not be noticed as such by the local population (Schafer 1994: 240). In the Alfama quarter of Lisbon, Portugal, there is an unfathomable amount of chirping and tweeting. Numerous canary birds sit in their tiny cages in the open windows of apartments and fill the air with a distinctive soundmark. Thinking about those kinds of signals might evoke memories of the past, as Schafer points out: "The hand-operated pump, also on the decline, now snaps into memory as a soundmark of my youth [...]" (Schafer 1994: 48). The last example illustrates that soundmarks are linked to places just as they are subjective memories. If soundmarks are unique sounds (Truax 2001: 67), it is because of the anchorage within the individual stream of consciousness and the associations that bind the sound to the listener. One the one hand, a soundmark is defined by its almost everlasting being (Schafer 1994: 239). On the other hand, there is the constant threat of extinction. Many soundmarks are no longer hearable, like the ones from medieval or Roman times. Because of their strong connection to the listener there seems to be a need in many to preserve the soundmarks by recording and storing them: "The soundmark, historical, and disappearing sounds, as well as the memories of those who can recall the soundscapes of the past, are all worthy of preservation and respect" (Truax 2001: 106). As described above acoustic ecology has its problems with preserving sounds. Like the recorded sound event that is transformed into a *soulless* sound object, soundmarks cannot be saved easily. Without being embedded into the soundscape with all other keynote sounds and signals, the soundmark loses its authenticity and degenerates into yet another sound object. Schafer and the World Soundscape Project nevertheless try to preserve many precious and not-so-precious sounds. How sound recording is approached will be discussed in the following section.

3.2 Sound Recording in the Field

There are various ways to record a soundscape. Most diverse approaches lead to numerous results that vary from technological accuracy to individual originality. Active listening can be considered the starting point of soundscape studies. "Thus our first question is 'What do we hear?'" (Gaver 1993: 3). The first attempts to sensitize listeners to sound were the listening sessions of Murray Schafer (Schafer 1969, 1970, 1994), the raw recordings of Barry Truax (Truax 1978, Truax 2002), and the *soundwalks* of Hildegard Westerkamp (Westerkamp 1974, 1988), all members of the Canadian World Soundscape Project. Decibel measurements help to learn more about certain qualities of sound, while participant observation, individual descriptions, and narrative interviews are focused on primarily. All these methods are considered sound recording, as experiencing sound and describing it allows us to understand individual relationship of the listener to the sound environment.

As section 3.2.2 will show, today individual findings become less important in favor of the more and more popular GIS technologies. Maps are created digitally, and visual design guides are invented to display sound qualities like energy and pressure level, frequency, and rhythm (Kornfeld 2008, Kornfeld, Schiewe, and Dykes 2011). Noise, especially traffic noise in urban areas, gains in importance. Research tries to present noise calculator software packages (Farcas and Sivertunb 2009), and "characteristic audio sequences [can be found] in common 3D city models" (Schiewe and Kornfeld 2009: 1). 3D replaces 2D mapping in digital terrain models, when terrain and buildings are displayed to calculate the impact of the physical environment on sound and its spatial reach (Arana et al. 2011). Even time is added to some simulations as a fourth dimension. Approaches try to cope with the fact of a constantly changing sound environment using a series of consecutive scenarios (Pamanikabud and Tansatcha 2010).

Decibel measurements support studies on noise with detailed recordings of the sound level at a given location during a certain amount of time. Values for day, evening, and night are separately measured in the field (Makarewicz and Zółtowski 2008). This most cost-intensive method is contrasted by the use of existing statistical data to calculate noise levels. Guidelines like those in the European Union help to maintain the balance between cost and accuracy (European Commission Working Group Assessment of Exposure to Noise (WG-AEN) 2006).

Standardized interviews have been used since the early soundscape studies of the World Soundscape Project. In questionnaires, researchers try to find out about the individual experience of the soundscape and people's preferences and feelings about sound (Raimbault, Lavandier, and Bérengier 2003). Interviews can take place in the field or under laboratory conditions (Irwin et al. 2011), and experts may be questioned (Raimbault and Dubois 2005), as well as passers-by and users of specific locations (Kang 2007). Sometimes sound samples are integrated into the questionnaires (Ge and Hokao 2005). Other studies take sound samples while conducting interviews (Schulte-Fortkamp and Genuit 2004). Narrative interviews are primarily used to pretest the questions and topics of a questionnaire, to learn about individual preferences with as much detail as possible. GIS-based models calculate noise levels (De Coensel et al. 2005), 3D models seem to be most sufficient for sound mapping (Krygier 1994), and field recording has been perfected regarding sample points and sound quality (Pamanikabud 2009, Tsai, Lin, and Chen 2009). To find out more about the impact of sound on the active listener and any person passively consuming it, individual focus also has to be covered in today's soundscape research (Iges 2000, Papadimitriou et al. 2009).

3.2.1 Methodology According to the World Soundscape Project

In 1973 the World Soundscape Project published what was to become the first popular study on urban sound: the *Vancouver Project*. Audio recordings and an accompanying booklet tell about Vancouver's most typical sounds in the 1970s. Methodologically, the participants of the project, like Murray Schafer, Barry Truax, and Hildegard Westerkamp, didn't have any blueprint for how to accomplish such an undertaking.

> [T]he intent was to document and re-present recordings of various sonic environments to the listener in order to foster awareness of sounds that are often ignored (Truax 2002: 5).

According to the scholars, sounds not noticeable at first have to be listened to very closely to attract listener's attention. So, the act of listening is of the highest importance. Schafer often started his classes by letting students sit still and listen to the environment. "The class had done this for four days running, ten minutes each day, chairs turned to the wall, receiving sound-messages" (Schafer 1969: 5). After that, students discussed their impressions and findings and thus started to reflect on sound and the soundscape. Before any other method can be applied, soundscape recording has to start with active listening to – not only hearing of – the most basic sound elements. "The first way is to cultivate the habit of listening. Listen carefully with seismographic delicacy to the sounds of the environment around you" (Schafer 1970: 3).

Listening to the sounds of a classroom might be a good start, but to experience the whole range of environmental sounds the listener has to be in the field. Hildegard Westerkamp describes walking the streets and paying attention to the multiplicity of sound sources as a "soundwalk" – an "excursion whose main purpose is listening to the environment" (Westerkamp 1974: n.p.) and which starts as attentive walking and intensive, but nondirectional, eavesdropping. (Werner 2006: 115) These walks are often recorded, though the listener's own walking sounds and breathing and other interferences have to be ignored when playing back the recording. Video might also be used to make deconstruction of the technically preserved soundscape easier. Taking pictures during a soundwalk can also help to clarify the audio material, although the sound of the release of the camera might be recorded, too.

Soundscape recording has been significantly changed over the last 40 years. What started with "a stereo Nagra IV-S and a pair of AKG condenser microphones"

(Truax 2002: 5) has turned into digital recording with notebook computers and "the 8-channel spatialization of environmental sound" (Truax 2002: 5). For their *Vancouver 1973* project the World Soundscape Project did 24-hour recordings of a certain site to produce a one-hour composition. As noted above, the presence of the listener, i.e., the researcher, can also be found in the recording, if s/he carries the recordings device(s). Even in the case of a fixed installation using a tripod, Werner points out that the mere existence of a recording device on site alters the recording, not least because of the physical presence of objects deflecting sound (2006: 75–6). In order to increase the fidelity of the recordings and bring them closer to natural hearing conditions, the World Soundscape Project used a binaural recording method, with "the recordist using his/her own pinnae with the microphones placed in the ears, or else artificial pinnae attached to the microphones themselves" (Truax 1978: n.p.).

Walking through the city and recording the soundscape can be accompanied by decibel measurements – another way of revealing the soundscape. Schafer did this while walking his students through residential areas (Schafer 1969: 13).

> The decibel is a unit for measuring the intensity of sound. It is used to express the relationship between the faintest sound man can hear and other sounds in the environment (Schafer 1970: 2).

"0 dB is defined as the THRESHOLD OF HEARING" (Truax 1978: n.p.). While 140 decibel hearing is possible, numbers as low as 90 can cause permanent damage to the human ear if exposure time is long enough. Table 3.1 shows decibel levels for different common sounds, including the maximum time of exposure before the possibility of hearing damage. The decibel scale is logarithmic, "so that an increase of 10 decibels mean a tenfold increase on sound intensity" (Schafer 1970: 2). Two sounds of the same decibel level combined do not double the loudness but rather increase the measured decibel level by three (Truax 1978: n.p.).

There is another way of measuring sound that is less technological. Schafer used descriptions of sounds heard as a way to learn more about the soundscape. Describing sound leads to classification, for example, human, nature, or technological sounds (see section 3.3). The naming of sound sources cannot only be applied to the environment of the listener but media as well. The sounds of objects and characters in a piece of art, e.g., in paintings or writings, can be analyzed. If a blacksmith is displayed swinging his hammer, this implies the sound of forging steel. Crows in the trees might evoke their croaking. People described as rushing through the streets make one think about shoes hitting the asphalt. The descriptions of all those sounds are based on the linguistic possibilities and the stream of consciousness of the researcher. "We only sampled at random, of course, but perhaps certain conclusions could be drawn" (Schafer 1969: 6).

Perhaps the most individualistic form of soundscape recording is based on participant observation. To write down what listeners experience while listening to

Table 3.1 Decibel Comparison Chart

	Decibel level (dB (A))	Distance (m/ft)	
Airplane	130	100	328
Fog horn	130	1,000	3,280
Live music concert	120	0	0
Ambulance siren	110	10	32
Disco	110	0	0
iPod (in-ear headphones)	109	0	0
Car horn	100	7	22
Lawnmower	100	1	3
Church bell	95	500	1,640
Bicycle bell	95	1	3
Motorbike	90	100	328
Barking dog	85	20	65
Traffic	80	5	16
Bus	80	15	50
Telephone	75	3	9
Playground	75	0	0
Sports field	75	30	100
Normal conversation	60	0	0
Birds	60	10	32
Quiet park	55	0	0
Quiet room	40	0	0
Whisper	30	2	6
Leaves	20	0	0

the soundscape is part of the World Soundscape Project's practices, and Hildegard Westerkamp goes one step further by focusing on her own relationship to sounds:

> The self's interaction with the acoustic environment becomes the focus of study. Since it is my intention to find out about other people's relationships to music-as-environment in a future case study, it seemed rather crucial to use myself as a case first (Westerkamp 1988: 148).

Other listeners are questioned in narrative interviews about their most liked and disliked sounds, the most typical or untypical sounds, or the loudest and most quiet ones. As with participant observation, the results are written down and analyzed later including classification.

To record the soundscape with pen and paper rather an audio device leads to the question of how sounds can be transformed from an audio wave to visible text.

According to Schafer, there are three different notation systems that can be applied to sound:

> We have three graphic notational systems available:
>
> - that of acoustics, by which the mechanical properties of sound may be exactly described on paper or a cathode-ray screen;
> - that of phonetics, by which human speech may be projected and analyzed;
> - musical notation, which permits the representation of certain sounds possessing 'musical' features (Schafer 1994: 123).

All three systems are descriptive and can display sound to an extent that simply talking about sound cannot cover. But notation is only applicable for certain sounds, like single sound events or a collage of clearly identifiable sounds. While music – Schafer calls it an organized form of sound (1994) – can be deconstructed into octaves and notes, the interplay of environmental sounds has to be recorded in a different way. Aerial projection is one method the World Soundscape Project uses. Mapping the soundscape with GIS software was not an option in the 1970s. Researchers collected sound data in the field, recording sound sources or sound heard in an area. The intensity of sounds and its distribution requires additional tools like decibel measurement devices. Thus, isobel contour maps can be created (Schafer 1994: 131). The recurrence of sounds is detected by counting and leads to another cartographic display of sound. The final maps are all based on data taken from the single sound events, which become sound objects when taken out of their contextual environment (see 1994: 274, Truax 1978: n.p.).

In the 1973 Vancouver Project researchers used recorded audio material to compose an artistic version of the soundscape. As stated above, from the 24 hours worth of recordings only one hour was finally used for the composition. The audio files were edited using a special technique:

> [T]ransparent editing and mixing of untransformed original recordings, selected for their aural interest and representativeness of the soundscape in question (Truax 2002: 5).

The selection of relevant passages within the recording was more important than enhancing the acoustic quality of the recording. Thus, a new, artificial soundscape is created that is supposed to represent the environmental pendant and activate listeners to become more aware of the sound that surrounds them. Active listening is promoted using well-known sounds that can be recognized easily. The researcher becomes sound designer or sound artist, who gets better insights into environmental sounds while recomposing the soundscape. Because of the individual character of soundscape composition many pieces contain programmatic messages, often related to the soundscape itself and topics like hearing or listening (Werner 2006: 135).

The environmental aspect of soundscape composition can be heard in the works of Hildegard Westerkamp. She collects sounds to recycle them for various purposes like movie scores, concerts, or sound installations. Werner describes how Westerkamp uses the sounds of the horns of ships to create "harbor music" and transforms the sound of breathing into relaxation compositions (see Werner 2006: 139). Like the Vancouver Project in 1973, soundscape compositions not only sensitize the listener to the acoustic environment, but it also serves as a way of communicating the researcher's message:

> [M]y own creative compositional work has, up to this point, occurred almost exclusively in sound studios. On the positive side, the studio environment has provided me with a little 'niche' where I could find my creative inner voice without interference from the surrounding social, cultural context. It allowed me to imagine, invent and use my fantasy freely. [...] In my electroacoustic compositions my inner voices speak and in that form I have been able to make them public (Westerkamp 1988: 133–4).

3.2.2 Data Acquisition

Acoustic ecology's individual approaches to sound recording from the 1970s have mostly been replaced by more quantitative techniques and modern GIS models. Listening – once described as the most fundamental method to explore and understand the soundscape – is not a part of soundscape research in most scientific disciplines anymore, while sound recording is still popular. Using real environmental data might be harder to collect than statistical values on sound (see below), but it is still unbeatable in terms of accuracy. Pretesting the soundscape of the research area is one way to get to know the acoustic environment and to single out possible sound events for recording. This can either be accomplished by soundwalking through the research area (see the empirical study in Lisbon in 3.4.1) or setting up microphones to later deconstruct the recording.

> [W]e performed pilot sound recordings to capture all available sound categories and to associate them with specific landscape characteristics (e.g. roads, construction, coastal zone) (Papadimitriou et al. 2009: 129).

After a possible pretest, recording is conducted with a precise setup taking into account technology, placement, and installation. Today, this technique is used especially to find out about noise issues in urban areas. Pamanikabud's (2009: 24) approach to recording noise data "on several locations along the side of motorway" is extremely precise if:

> data are collected using precision sound level meters set on tri-pod with the height of 1.20 m from ground surface. These sound level meters are located

> randomly at the distances 1m further from the edge of near side frontage road (Pamanikabud 2009: 25).

Employing the right number of recording points is important for getting significant information about the local soundscape. Arana et al. come to the conclusion "that mapping with a 0.5 m degree of accuracy in elevation is sufficient for acoustic simulation" (2011: 522). For studies on traffic noise this short distance is not needed. The study of Tsai et al., for example:

> established 345 temporary noise monitoring stations in the Tainan metropolitan area […]. To select the locations of measuring points, the Tainan metropolitan area was divided into standard grids with size of 500 m × 500 m (2009: 967).

Papadimitriou et al. use fewer stations but a tighter grid of 350m distances. The purpose of positioning microphones this way "could ensure the recording of a great variety of sounds from different sources, and allows the interpolation of the collected information" (2009: 129).

The interval is of equal importance. To get a sufficient sample for sounds like traffic noise, different times and days, even months and seasons, are of importance. Traffic density varies between day and night, weekdays and weekends, and summer and winter (see Ge and Hokao 2005: 458, Tsai, Lin, and Chen 2009: 964). To process traffic noise data the averaging of decibel values is common practice. The average sound level L is usually divided into three sections: day, evening, and night (Makarewicz and Żółtowski 2008: 3568). The exact time when the evening ends, the night starts, and the day begins depends on the given situation. The "[d]efault duration defined in the END [environmental noise directive of the European Union]: day (12h: 07:00 – 19:00), evening (4h: 19:00–23:00), night (8h: 23:00–07:00)" (European Commission Working Group Assessment of Exposure to Noise (WG-AEN) 2006: 63). The average sound level of day, evening, and night combined is named L_{den}.

Problems recording sound generally do not occur because of technology but the character of the environmental soundscape as a cacophonic mixture of keynotes, signals, and soundmarks. The interplay of sounds and their sources with the physical environment constantly alters the single sound.

> [M]uch of the sound that reaches us from a source has reflected off various other objects in the environment, which colour the spectrum of reflected sound just as light is coloured by the surfaces it strikes (Gaver 1993: 4).

This shows again how crucial pretesting can be. A pretest can reveal the importance of a sound for the overall soundscape and its significance among the other sounds. Background sounds may be singled out and be considered influencing factors for the sounds that get recorded in the final study (Pamanikabud 2009: 129). In the end, the act of recording itself alters the sound (see above). It does not produce a copy of reality but reflects it under certain circumstances (like the positioning of microphones). The result "is a kind of sound mirage: an acoustic image which may aim to represent the reality in question, while actually presenting a distorted reflection which, at best, captures the aroma of the original" (Iges 2000: n.p.). This, of course, refers to the nature of sound that is so important to acoustic ecology. For traffic noise measurements and GIS models the difference between the natural sound event and the preserved sound object is inexistent.

In the *Good Practice Guide for Strategic Noise Mapping and the Production of Associated Data on Noise Exposure* the European Union describes different possibilities for acquiring sound data (2006). For various measurements like the sound of heavy vehicles, road surface type, or speed fluctuations at road junctions a toolkit is provided to the member states. On-site recording of the soundscape is always an option, but statistical solutions are offered, too. Generally, field recordings are the most complex, most expensive, but also most accurate way of acquiring data. The example of traffic flow illustrates the procedure. If no data of the traffic flow at a certain crossing, for example, is available, researchers have the option to "[m]ake traffic counts for each of the three periods: daytime, evening and night time" (European Commission Working Group Assessment of Exposure to Noise (WG-AEN) 2006: 59). Complexity is very high, increasing with the number of vehicles and their speed. This method is extremely expensive, as a certain number of people have to count for a long period of time, while errors can be avoided best if several researchers have the same area to cover. The final data set will have a tolerance of at most half a decibel. If a city council has to create a noise exposure map covering all major roads, a complete counting is not an option. Instead, it is possible to "[s]elect sample roads and do traffic counts there; extrapolate to other roads of same type" (European Commission Working Group Assessment of Exposure to Noise (WG-AEN) 2006: 59). The complexity stays as high as before, but cost might come down a little, depending on the number of staff. Accuracy goes down to plus or minus two decibels. A much easier and more cost-efficient way to get sound data is to "[u]se official [or other] traffic flow data for typical road types" in the country or region (European Commission Working Group Assessment of Exposure to Noise (WG-AEN) 2006: 59). In terms of accuracy – there can be anomalies of up to four decibels – this method delivers the same results as using the default values of the European Commission, provided in the *Good Practice Guide*. The only complexity left is to add up the different road types like service roads and collecting roads, and to apply standardized values for day, evening, and night traffic to them (see European Commission Working Group Assessment of Exposure to Noise (WG-AEN) 2006: 59). High costs of individual counting lead to the fact that documents like noise exposure maps of urban areas are generally produced using statistical data. For every study "[t]here is a need to compromise between the accuracy in the determination of L_{den} and limited resources of input data, the measuring equipment, and time" (Makarewicz and Zółtowski 2008: 3568).

Recorded sound data has to be processed to draw conclusions out of it. One core geographic way to do this is mapping. There are various possibilities to approach sound mapping and it seems that it is far from being perfect (Gaver 1993). Certain qualities can be visualized, like sources, reach, pitch, or decibel level. But at least one element cannot be shown in traditional maps: time. As the soundscape consists of many interacting sound events the as-is state is changing constantly. Noise mapping along a railroad track, for example, depends tremendously on the sound caused by passing-by trains. Average values like L_{den} are not sufficient, as the highest decibel peaks are of importance. If a "traffic noise simulation model in 3D form is applied on a GIS system" (Pamanikabud and Tansatcha 2010: 1185) free-flow sound can be investigated. "[C]olor-enhanced noise contours on the ground and building surfaces" can be added to make changes visible (Pamanikabud and Tansatcha 2010: 1193).

In static models this problem can partly be solved with a time series of different scenarios. Calculations and visualization are mostly done in GIS software. This helpful tool was not available to the World Soundscape Project in the 1970s, but GIS software researchers today are able to produce several mapping concepts for one scenario using high-performance computers (Kornfeld 2008): "According to our encodings, we build a high-level design guide for the visualization of sound" (Kornfeld, Schiewe, and Dykes 2011: 26). Visualization varies from point signatures for sources or recording points to lines, pixels, and areas for reach, loudness, and other sound features. "The natural shape of sound is a wave traveling through air. Due to large scales, this is abstracted to a simple geometric shape of sound" (Kornfeld, Schiewe, and Dykes 2011: 17). A wave lying over an urban area is not displayable in a 2D environment. Auralization is one way to solve the problem. "[M]ethods of traditional, digital and multimedia cartography as well as the integration of characteristic audio sequences in common 3D city models (auralization) are intended to be incorporated into the urban sound cartography" (Schiewe and Kornfeld 2009: 2496). Different color-coding helps deconstruct the soundscape. Single sound sources within a wave (or similar representation) show how the soundscape is build up. Sources are categorized into traffic and economy, human activity, and nature sounds. Color gradients signify sound energy and sound pressure level. Spatial reach and rhythm are displayed with colored areas (Kornfeld, Schiewe, and Dykes 2011: 26).

The focus of GIS-based approaches to mapping sound varies, from displaying large areas like cities or agglomerations to the description of single streets or public green spaces. Depending on the study, new software solutions are presented (i.a., De Coensel et al. 2005, Farcas and Sivertunb 2009, Kornfeld 2008) that all use slightly different color-coding and categorization systems. Tsai et al., for example, work with a green-to-red scale symbolizing "[w]inter noise distribution in Tainan City" during morning, afternoon, and evening hours (Tsai, Lin, and Chen 2009: 967). To display "[s]ummer violations of noise regulations" the authors decide to move from yellow to red and skip the green hues because the negative connotation of noise is in conflict with the positive connotation of the color green (Tsai, Lin, and Chen 2009: 968).

A 3D model contains as many physical objects as possible. Not only the terrain but buildings and other structures are worked into the simulation (Arana et al. 2011: 522). The results of the added noise data finally depend on the accuracy and the number of recording points. "With a vertical resolution of 0.2 m, the results are virtually identical to those obtained for a 0-tolerance. From our view point, mapping with 0.5 m in vertical accuracy is sufficient for acoustic simulation" (Arana et al. 2011: 526).

As discussed earlier (see 1.4), 3D models seem to be more effective for sound mapping (Krygier 1994: 149). But even though GIS offers enormous possibilities for visualizing sound in maps:

> the task of actually describing and depicting the soundscape is more complicated than noise mapping, since as the communication model implied by Truax [...], subjectivity is inserted to the observation (Papadimitriou et al. 2009: 127).

Research on sound in disciplines like social geography focuses on the individual and common perception of sound, and not so much on its physiological extension. This mean, that 3D models that include terrain, buildings, and vertical sound emissions, and even animated 4D models with the capability to display changes over time, are not primarily of interest. Traditional 2D cartography with data on sources and reach of sound seems to be appropriate. One passer-by might be taller than another one and thus receive sounds differently. But the possible discrepancies caused by a slightly different sound mixture can be neglected. The highly varying experiences of sound that are based on individuality, social context, and one's own stream of consciousness are far more important.

Qualitative and quantitative interviews offer the chance to integrate an individual aspect into sound recording. That is why some studies conduct interviews while recording audio data in the field simultaneously (Schulte-Fortkamp and Genuit 2004: 2496). Raimbault et al., for example, describe "a cross analysis of survey data and acoustic parameters, describing typical ambient urban sound environment" (2003: 1242). At every audio recording location people were asked questions about the local soundscape, using a questionnaire. Qualitative narrative Interviews had been conduced first to build up a base for the quantitative questions (Raimbault, Lavandier, and Bérengier 2003: 1244). It seems that two types of questions were most heavily used to find out more about the respondents' soundscape experiences:

> The semantic differential profiles of soundscape are useful in helping us to understand the perceptual nature of the sound environment of urban street areas, and to create deeper context of soundscape (Ge and Hokao 2005: 465).

Aside from the semantic differential (also see Raimbault, Lavandier, and Bérengier 2003: 1245), 5-point and 7-point scale questions were named to help quantify one's personal feelings toward a sound, regarding likability, loudness, and familiarity. "The interviewees were also asked to classify at least three sounds as 'favourite' (F), 'neither favourite nor annoying' (N), or 'annoying' (A)"

(Kang 2007: 56). The sounds in this example were prerecorded in the neighborhood and replayed to the respondents. To ensure identical conditions during questioning, some studies conducted interviews in a laboratory (Irwin et al. 2011). As the first pretest in Lisbon (see 3.4.1) will show, it is also possible to randomly question passers-by all over a predefined area. Thus, the group of respondents can differ from a random selection of people on site to recruited sound experts, divided by age, sex, and knowledge.

Narrative Interviews are less common in today's soundscape research. Sometimes used as a pretest to cover possible questions for the quantitative questionnaire, it seems that higher levels of individual answers from the narrative interview do not work well in combination with audio recording and statistical data. However, to gain initial background information on a certain soundscape, narrative interviews can be of value. In the case of Raimbault and Dubois:

> [i]nterviews of a representative panel of French planners were therefore analyzed in order to understand their awareness concerning the place and utility of sound ambient environments among urban projects (2005: 341).

Gathered data is analyzed digitally using software like MAXQDA, atlas.ti, and Nomino. The combination of interviews, recordings, and statistical data can not only ensure accuracy but integrate the individual elements of sound perception that were once the starting point for World Soundscape Project soundscape studies.

3.3 Classifying Sounds in Different Ways

In the field, recording techniques vary, from descriptions to audio recordings. The documentation of thoughts and audio data always includes categorization and classification. For the World Soundscape Project, Schafer describes a basic trichotomy of nature sounds, human sounds, and electric or mechanical sounds (Schafer 1969). Nature sounds are defined as the most pleasing, mechanical sounds as the most annoying (Schafer 1970). A more detailed analysis distinguishes between acoustics, psychoacoustics, meaning, and aesthetic qualities (Schafer 1994). The soundscape as a whole is always part of the process. Sound objects, detached from the surrounding acoustic environment, lose meaning. After being recorded, sound can no longer reflect all qualities and thus fails to describe a soundscape completely. Sonography, the "art of soundscape notation", is limited (Truax 1978: n.p.).

Depending on research focus, most recent studies use only some of the qualities introduced by Schafer. Gaver highlights hearable features of a sound event with his classification of vibrating solids, aerodynamic events, and liquid solids (1993). Both Servigne et al. and Lebiedowska predetermine certain sounds as noise, when they focus on traffic, road works, factories, people (Servigne et al. 1999), background noise, mechanical equipment, human activities, noises of nature, human presence, and speech and communication (Lebiedowska 2005). Anthropophony, biophony, and geophony (Papadimitriou et al. 2009) replace the basic trichotomy of nature, human, and mechanical sounds. Intensity of a sound event is also integrated into the analysis.

Schiewe and Kornfeld highlight subjective evaluation as a base for aural perception. In addition to acoustic and psychoacoustic qualities of a sound event, individual perception is also considered (2009). That such classification does not have to rely on empirical field work is shown in the works of Yu and Kang, who add instrumental sounds to Schafer's three basic categories when considering the audibility of single sounds (2010). Also, working with prerecorded sound samples, Rychtáriková and Vermeir classify sound according to acoustic and psychoacoustic features (2013). Overall, the classification of sound events always depends on scientific context and the aim of the study.

3.3.1 Acoustic Ecology's Classification

With listening, soundwalking, and sound descriptions, the World Soundscape Project approached the study of the soundscape in the late 1960s and early 1970s. All findings had to be recorded and classified to place them in the context of the entirety of sounds. "The question is, which types of notation will be most helpful in these pursuits? At present there can be no grand solutions to this problem, for research is only beginning" (Schafer 1994: 131). According to the Project, sonography itself might be a form of classification, depending on the approach to "register the geographic distribution of sound events" (Truax 1978: n.p.). Schafer tries different ways to deconstruct the soundscape, in which the researcher's role becomes a determining factor. It takes not only the quantitative data of sound events but also individual evaluation to create a proper classification system: "It will only be accomplished by a new generation of artist-scientists trained in acoustic ecology and acoustic design" (Schafer 1994: 134). The result is believed to be comparable to an object of surrealistic art that "also depends on bringing together incongruous or anachronistic facts, which nevertheless somehow snap together to illuminate new relationships" (Schafer 1994: 133).

Most of the time the World Soundscape Project analyzes the single sound events the soundscape consists of. Acoustic attributes like loudness and frequency are important but not preferential for classification. Context and, most of all, meaning build the base for determining sounds. Sound objects – described above as the recorded equivalent to the sound event – are lacking both. "[A]wareness of sound symbolism will be useful in trying to decide which sounds we want to preserve in the world and which we want to suppress" (Schafer 1970: 15). The answer to this question of preservable sounds lies in the individual interpretation of the researcher, social beliefs, and time. The latter refers to a constantly changing urban soundscape that is expanded through the sounds of industry's heavy machinery, Fordism's automotive engines, and turn-of-the-millennium mobile phones and Internet communication devices. "[A]fter the Industrial

Revolution, mechanical sounds drowned out both human and natural sounds with their ubiquitous buzz and whirr" (Schafer 1969: 6). Onomatopoeic words like *buzz* and *whirr* are relatively neutral descriptions for unnatural sounds and do not point to a classification, while the category of *mechanical sounds* already is one. Before Schafer describes certain ways of classifying sound events he determines three major sound categories: nature sounds, human sounds, and electric or mechanical sounds (Schafer 1969: 5–6). This trichotomy is often used in acoustic ecologies and beyond, while explanations are generally missing. From a social geographic perspective, classifying sound based on its appearance or perception is suitable, especially in the constructivist approach of viability – i.e., classifications "prove adequate in the contexts in which they were created" (von Glasersfeld 1995: 7).

> Like all techniques of analysis, this can only be justified if it leads to the improvement of perception, judgment and invention (Schafer 1994: 133).

In Schafer's perspective, nature creates the most positive sounds for the human ear. Wind shaking the leaves of trees, birds singing, and water flowing are examples. That nature sounds "have endured the longest" is quite obvious, but they are also described as being "mostly pleasing to man" (Schafer 1970: 14). The category of human sounds consists of people's voices, the rustling of cloth, and shoes hitting the ground, etc. He does not give a description for the words *pleasing* or *unpleasant*, in contrast to electric and mechanical sounds. Technology – at least in Schafer's 1970s perspective – has a positive and future-oriented connotation to most people. For him, technology sounds are far less pleasing than nature sounds (from here I will use the term 'technology sound' to describe both mechanical and electric sounds). Noise is often named in the context of urban sounds that are caused by mechanical or electric sources. In addition to getting people to listen to their environment, the purpose of the World Soundscape Project is to sensitize listeners to perceive sounds in a certain way:

> It will only be possible to discriminate against the sounds of a conscienceless technology and muffle its plangent voice when its sound-symbolism changes in the imagination of sufficient numbers of men (Schafer 1970: 15).

In addition, Schafer spots four ways to classify single soundscape elements, depending on context and scientific discipline:

> The physicist and engineer study acoustics; the psychologist and physiologist study psychoacoustics; the linguist and communications specialist study semantics, while to the poet and composer is left the domain of aesthetics (Schafer 1994: 148).

Corresponding to these four approaches, sound can be classified:

> according to their physical characteristics (acoustics) or the way in which they are perceived (psychoacoustics); according to their function and meaning (semiotics and semantics); or according to their emotional or affective qualities (aesthetics) (Schafer 1994: 133).

Physical characteristics can be recorded quantitatively, thereby reducing the impact of individuality. Data like *frequency* and *duration* of a sound can be measured objectively. A sound starts with the *attack*, while *body* describes the time during which the sound exists until its *decay*. A card for each sound is produced that displays all characteristics, which can be analyzed later. Psychoacoustic information adds to the acoustic information. Observations about basic conditions during recording, possible fluctuations of the sound, and external interferences like wind that might influence the original sound, are also named on the card (Schafer 1994: 136).

While acoustic characteristics are considered objective data, "no sound has objective meaning, and the observer will have specific cultural attitudes toward the subject" (Schafer 1994: 137). Semiotics and semantics can change, which makes it difficult to categorize sounds and separate them from each other. However, Schafer offers a way to deconstruct the soundscape according to the meaning of sounds via "their referential aspect" (Schafer 1994: 137). This aspect refers to the first categorization of sound events as nature, human, and technology sounds. So, if the sound of a bird singing is recorded, it belongs to nature and will most likely have a positive effect on the listener, while the roaring engine of a motor bike, as a technology sound, will most likely engender disturbance for the same person. If the listener is a fan of motorbikes, the latter example seems to be, at least, problematic. How listeners perceive and associate a certain sound event cannot be scientifically verified. Whenever soundscape classification aims at producing results that lead to an "improvement of perception, judgment and invention" (Schafer 1994: 133) meaning is inscribed in advance to categories like nature and technology. This may help sensitize listeners, but it puts perception into a social corset of stereotypes and generalization. Schafer's referential aspects are not limited to nature, human, and technology sounds. Aspects like *sound and society* or *quiet and silence* are also discussed. The former indicates the problem of socially founded cliché. The latter makes clear that the named aspects are categories used very generally. As Chapter 4.5 will show, silence is highly dependent on individual perception and can mean various things to the listener (see also 2.4). Thus, referential aspects can only be a starting point for an analysis that locates recorded sounds within common meaning and understanding. Aesthetic qualities – the fourth way of classifying sounds – is "probably the hardest of all types of classification" (Schafer 1994: 146). Labels like *ugly* and *beautiful* are relative terms that, like meaning, strongly depend on the listener's sound experience. "Sounds affect individuals differently and a single sound will often stimulate [...] a wide assortment of reactions" (Schafer 1994: 146).

The ascription and meaning of sound are always dependent on its context. Without it, sound events degenerate into meaningless sound objects. After a sound is recorded, it loses its embedment in the whole soundscape, with all its connotations and interrelations. From Schafer's point of view, context is highly important and has to be taken into account in order for researchers to learn more about environmental sounds. Sound is more than its frequency, decibels, and amplitude modulation.

3.3.2 Other Possibilities to Classify Sound

Starting with the classification system of Schafer and the World Soundscape Project, many authors have developed a slightly different way of categorizing sounds. The research focus is always the determining factor. For Gaver, sound sources are most important to explore. He assumes that "[e]ach source of sound involves an interaction of materials" (Gaver 1993: 4). When two objects touch each other, like wheels running on asphalt, the contact causes a sound. All those sounds can be experienced by the listener – that is why Gaver asks, "What do we hear" (1993). Hearing a sound at the same time means that empirical data can be recorded for further analysis.

> We can hear an approaching automobile, its size and speed. We can hear where it is, and how fast it is approaching. And we can hear the narrow, echoing walls of the alley it's driving along. These are the phenomena of concern to an ecological approach to perception (Gaver 1993: 5).

To categorize these sounds Gaver introduces three different categories: vibrating solids, aerodynamic events, and liquid solids. Examples for the first would be a door closing or walking of shoes. The second category includes sounds "caused by the direct introduction and modification of atmospheric pressure differences from some source. The simplest aerodynamic sound is exemplified by an exploding balloon" (Gaver 1993: 8). The third category covers sounds caused by an object falling into the water or any other liquid. "[D]ripping and splashing" are given examples (Gaver 1993: 9). Additional levels of classification include the "nature of the material", its configuration, and "according to whether they belong to the causal interaction" (Gaver 1993: 10).

Servigne et al. isolate a certain type of sounds that already is a classification in itself: noise. As discussed earlier (see 2.1) there are various sources for noise. The authors identify four categories that are based on the sound source: "Noises are caused by traffic, road works, factories, and also by people" (1999: 262). A similar focus on noise comes from Lebiedowska (2005). Like in Servigne et al., a detailed definition for noise is missing (various approaches are discussed in 4.2). The classification in the study is based on the intensity of noise in urbanized areas. Decibel measurements are conducted to later categorize locations from very quiet areas to very loud ones. Traffic noise is excluded from the soundscape and investigated in detail (Lebiedowska 2005: 342). Intensity can also be combined with an approach evocative of the studies of the World Soundscape Project. Human activities, sounds "produced by living organisms", and those of nature are explored by Papadimitriou et al. (2009: 129). Technology or mechanical sounds are missing in this description, and the differentiation between the second and the third category is not always clear to define. The sound caused by wind belongs to the "natural elements" category, but the rustling of leaves in the wind could be ascribed to the tree as a living organism. On the second classification level intensity is "scored in a three-level scale ranging from one (less intense sound) to three (more intense sound)" (Papadimitriou et al. 2009: 129). Like Lebiedowska, foreground and background sounds are separated from each other. Thereby, the group of background sounds differs from foreground sounds only because of its distance from the recording site:

> [T]he foreground sounds refer to those produced instantaneously and sharply near the sampling site (e.g. birds, insects) while the background sounds refer to those produced far away from the sampling site and originate from the whole surrounding landscape (e.g. traffic noise, long distance sound sea waves) (Papadimitriou et al. 2009: 129).

A second reason to differentiate between foreground and background sounds is the difference between those being produced "instantaneously and sharply" and those originating "from the whole surrounding landscape" (Papadimitriou et al. 2009: 129). One crucial factor is the distance from the recording device. At another recording site, traffic sound that is indefinable most of the time is deconstructed into sound events from certain car engines, horns of busses, and wheels of motorbikes.

Schiewe and Kornfeld (2009) focus less on the position and more on the acoustic and psychoacoustic qualities of a sound source. Subjective evaluation – aside from the unique location where sound is experienced – is tied to social meaning and the individual stream of consciousness is the basis for the classification. Determining factors include "technical acoustic variables such as sound pressure, frequency, time flow and sound propagation, direction and distance" (Schiewe and Kornfeld 2009: 1). In their sound-mapping catalogue (described in 3.2.2) various interdisciplinary classifications are used to categorize different elements. Geographic indicators deriving from urban geography or climatology are considered, as are those from soundscape research like lo-fi and hi-fi (see Westerkamp 1988: 6). Acoustics tell researchers about decibels, velocity, and intensity, while psychoacoustics inform them about loudness, pitch, and rhythm (Schiewe and Kornfeld 2009: 3).

Sound classification does not require fieldwork. Some of the most recent studies use sound samples to find out more about the soundscape and listeners' preferences. Yu and Kang adopt Schafer's trichotomy of natural sounds, human sounds, and mechanical sounds and add instrumental sounds as a fourth category

(Yu and Kang 2010: 625). In addition to this classification, the presence of sounds is tested in audio recordings played back to a group of respondents in a laboratory experiment. In contrast to field studies, certain sounds – e.g., traffic and birds, or a fountain and construction sounds – are combined and simultaneously played back. The former directive to always include the acoustic environment of a sound event into its classification and analysis is replaced by a focus on acoustic and psychoacoustic, even aesthetic, qualities. Meaning is less important, as the study of Rychtáriková and Vermeir shows.

> Soundscape categories are established as a result of an automatic clustering algorithm based on multi-parameter analysis by 13 acoustical parameters used as similarity measures, on a large set of sound recordings (2013: 240).

But semiotics and semantics are important features that are needed to fully understand the soundscape's impact on the listener.

3.4 Empirical Studies in Lisbon, London, and Austin

Sound fieldwork can be conducted using a variety of approaches. As the studies described above have shown, method is highly dependent on scientific discipline and the study itself. The present research seeks access to soundscape studies by using some of the methods described above and by offering some new techniques to approach urban sound. The most basic requirement is to record the soundscape. For that purpose, Schafer's trichotomy categorization – keynote, signals, and soundmarks – was first tested in a field study in Lisbon, Portugal. During a three-day period, the soundscape was recorded in soundwalks as suggested by Westerkamp (1974), following the creative approach by Werner (2006). A focus on sound events and the extension of fieldwork using a standardized questionnaire made it possible to learn about the deconstruction and classification of the soundscape.

The recording of sound events was part of a subsequent study in London, England. The number of research sites was extended (from nine in Lisbon to 12), and sound classification started with another approach from the World Soundscape Project. The classification of sound events as nature, human, or technology sounds made possible a vital discussion on the individual impact on soundscape fieldwork. It was determined that the perception of the researcher might collide with established classification concepts. For the study, the sound of leaves on a tree was classified as an inanimate nature sound, a concept most biologists might find offensive. In the context of human perception and the impact of sound on the urban dweller a more detailed classification system was developed. At the completion of the study, a question about typical sounds for certain areas or locations arose, leading to a third field study, conducted in Austin, Texas.

The Austin study was aimed at finding a way to describe the soundscape of a specific area, district, or quarter of a city. What are the most typical sounds at a certain location? The groundwork laid by the Lisbon and London studies helped to develop a method to answer this question. Soundwalks were a valuable method for approaching an urban area. Additional recordings and textual and photographic documentation delivered insights into soundscapes of a predefined area. At five research sites, from the dense city center to quiet urban green spaces, data was gathered for *sound point recording*. This new technique offers a way to describe the soundscape in detail. With the results of the Austin study, various research questions could be generated and an empirical fundament was established from which to proceed with the social geographic research projects described in Chapter 4.

3.4.1 Deconstructing the Soundscape – Lisbon

In the 1994 German movie *Lisbon Story*, Phillip Winter, a sound recordist, tries to help out his friend Friedrich Monroe, who is filming a documentary in the Portuguese capital. Because Monroe has vanished by the time Winter arrives, the recordist walks the streets of Lisbon to randomly sample sound events for the left-behind silent movie clips. The result of the more or less aimless soundwalk are recordings Werner describes in his "Soundcape-Dialog":

> Tram, birds, river – these are the major sound streams a listening flâneur in Lisbon experiences. And the traffic, the traffic that shall never end (2006: 105, translated from German).

According to the World Soundscape Project, walking around and listening to sound is no aimless practice. Werner – like Winter in the movie – approaches Lisbon's soundscape by soundwalking. "A soundwalk is any excursion whose main purpose is listening to the environment. It is exposing our ears to every sound around us no matter where we are" (Westerkamp 1974: n.p.). Werner extends the method by adding individual perception to the analysis criteria. What he calls the *MetaSon* of Lisbon is "experienced, perceived sound, that includes everything hearable, both real and imaginary. The soundscape, not how it is, but how it is perceived. MetaSon suggests a method of listening for urban environments" (2006: 110).

In my own study in Lisbon, I tested the soundwalk as a method for urban geographic sound research. As shown above, *describing* is one basic technique for the World Soundscape Project. Thus, starting with a description of the Lisbon soundscape seems to be adequate. In the first phase, a group of 18 students were asked to walk the city of Lisbon and perform a soundwalk. To add to their individual descriptions, audio recordings of the walks were made. As in Schafer's empirical studies (1994), the participating researchers were sensitized for listening and using methods of acoustic ecology. In pairs the students walked different parts of Lisbon for three to four hours to learn about the local soundscape. Mainly Baixa, the city center, and close-by quarters like Alfama and Chiado were investigated.

Heavy traffic and a high number of pedestrians dominated the descriptions, and some of the typical sounds depicted by Werner (2006) – especially the traditional tram – were named. Each research group not only recorded their soundwalk but took pictures of dominant sound sources, like canaries in the Alfama, talking people in the Baixa, and the tram in various other locations. A description of a representative soundwalk notes:

> We start our soundwalk at a little park North of Rua Alfândega and Rua Arameiros. Some people walk by, eating. Others sit outside the restaurants and have lunch. Some people talk. I can hear the sound of shoes passing by. They walk rather fast, probably on their way back to the Baixa after lunch-break. There is some traffic on Rua Alfândega. Mostly cars pass by, some vans, some trucks. A group of tourists stand at a corner. They talk in a foreign language (probably spanish). At a public parking lot two cars try to find a place to park. Behind the cars I can see the river Tejo and a dock (Doca da Marinha). Traffic sounds and some people walking by dominate the soundscape. The Tejo is too far away to be heard. At the National Industrial Property Institute (INPI) we decide to turn left into an archway to find a more silent acoustic environment. Under the archway reverberation is high and the sound of our own shoes and the clicking of the camera are the most dominant sounds. On the other side we enter Rua São João da Praça and turn right. We walk a little uphill and pass the Porta de Alfama. Church bells begin to ring very loud (probably from close by cathedral Sé de Lisboa). There are almost no cars passing by, but I can still hear traffic sounds. From uphill comes the ringing of an old tram. A dog is barking. Before we turn left to enter the Rua da Adiça, a motorbike passes us by. Now it is getting quieter. At the end of Rua da Adiça we continue our walk through a small pathway that passes two closely spaced apartment buildings. Again, there is more reverberation. I can hear multiple canary birds that get louder the further I go. I woman passes me by and she greets me ("Bom dia"). I do not respond because of the recording and enter a triangular open space. A lot of sounds can be heard here. Multiple canary birds, whose cages I can see in the open windows of the apartment buildings. The clicking of the camera is there, as always. People (I would guess two man) are yelling at each other. I cannot see them, but it sounds angry and emotional. A boy walks by, than two older women. There is a light chop or ripple of water. As I walk across the triangle I can see a fountain, where water falls into a stony basin. I try to record only the water as I step really close. The Rua Norberto de Araújo goes uphill and is rather quiet. I cannot locate from where I hear some traffic sounds. It is so quiet that the wind shaking some leaves and flowers should be heard in the recording. Shortly before the street turns left there is a small green space between two buildings on the right. From there I can see the Tejo and record a container ship's horn that floats on the river. (If there are more ships to be heard in the recording I cannot say. Maybe the visual impression supported me hearing the ship.) A few stairs go up the street and we come to a highly frequented spot: a place with cafés and restaurants on Rua

> Limoeiro and Rua de São Tomé. Trams pass by every two minutes and the sound of the wheels on the tracks is very significant. A lot of younger people sit in the café and walk the place, many of them heavily talking. There is a man playing the guitar. The sounds of honking cars, ringing trams, laughing people, music, and shoes walking mix up. We stop our soundwalk at a balustrade above Beco de Santa Helena (Wissmann 2010, translated from German).

As this example shows, describing a soundwalk forces the researcher to focus on sound within the urban landscape (the route of the soundwalk is displayed on Figure 3.1). Some passages indicate that visual impressions are predominant for perceiving the environment. More quiet sounds like those caused by wind may always be there but are only noted whenever louder sound sources are missing. Following the soundwalks, in a group session individual impressions were discussed. Presentation of the photographed sound sources and the environment helped to prepare for the second phase of the study.

In the second phase single sound events were recorded. The aim was to deconstruct the soundscape further by dividing it into the three main categories of keynote, signal, and soundmark. To record keynote and sound signals the previous soundwalks provided sufficient information. Soundmarks are not as easy to detect. To find out about sounds that locals as well as foreigners would describe as special for Lisbon, their community, or a certain quarter, a questionnaire with 10 questions was designed. Two hundred and forty-seven standardized, quantitative interviews were conducted in the central quarters of Baixa, Rossio, Chiado, and Alfama. The most common sound noted for Lisbon was the tram (34 percent), followed by cars and traffic (25 percent), and Fado (21 percent), the traditional portuguese music style. Asked which district or part of Lisbon sounded unique, respondents answered Alfama (35 percent) most often. Bairro Alto (24 percent), Baixa (22 percent), and the river Tejo (17 percent) received the next-highest number of mentions, followed by Chiado (11 percent) and Belem (6 percent). Graça (3 percent) and Rossio (2 percent) were mentioned least. The tag cloud in Figure 3.2 illustrates the results, and raises a question about what sounds in particular were considered "special" in the areas mentioned.

Within the so-called *sound cloud*, the closer a sound (source) is placed to a district the more important it is for the ascribed soundscape of that district. *Localities*, for example, was only mentioned in association with Alfama and Bairro Alto. Thus, it was placed between those two. Because this sound was more important (i.e., received more mentions) for the Bairro Alto soundscape, it was positioned next to the district's name.

In anticipation of the subsequent study in London, questions according to the common trichotomy of nature, human, and technology sounds were integrated (Schafer 1994). The respondents recalled mostly technology sounds related to traffic and transport (84 percent), followed by almost equally named human sounds (70 percent) and nature sounds (69 percent). In reference to Yu and Kang (2010) a question on music was integrated. It's little surprisingly that 94 percent of the respondents recall hearing music typical for Lisbon or that Fado was the

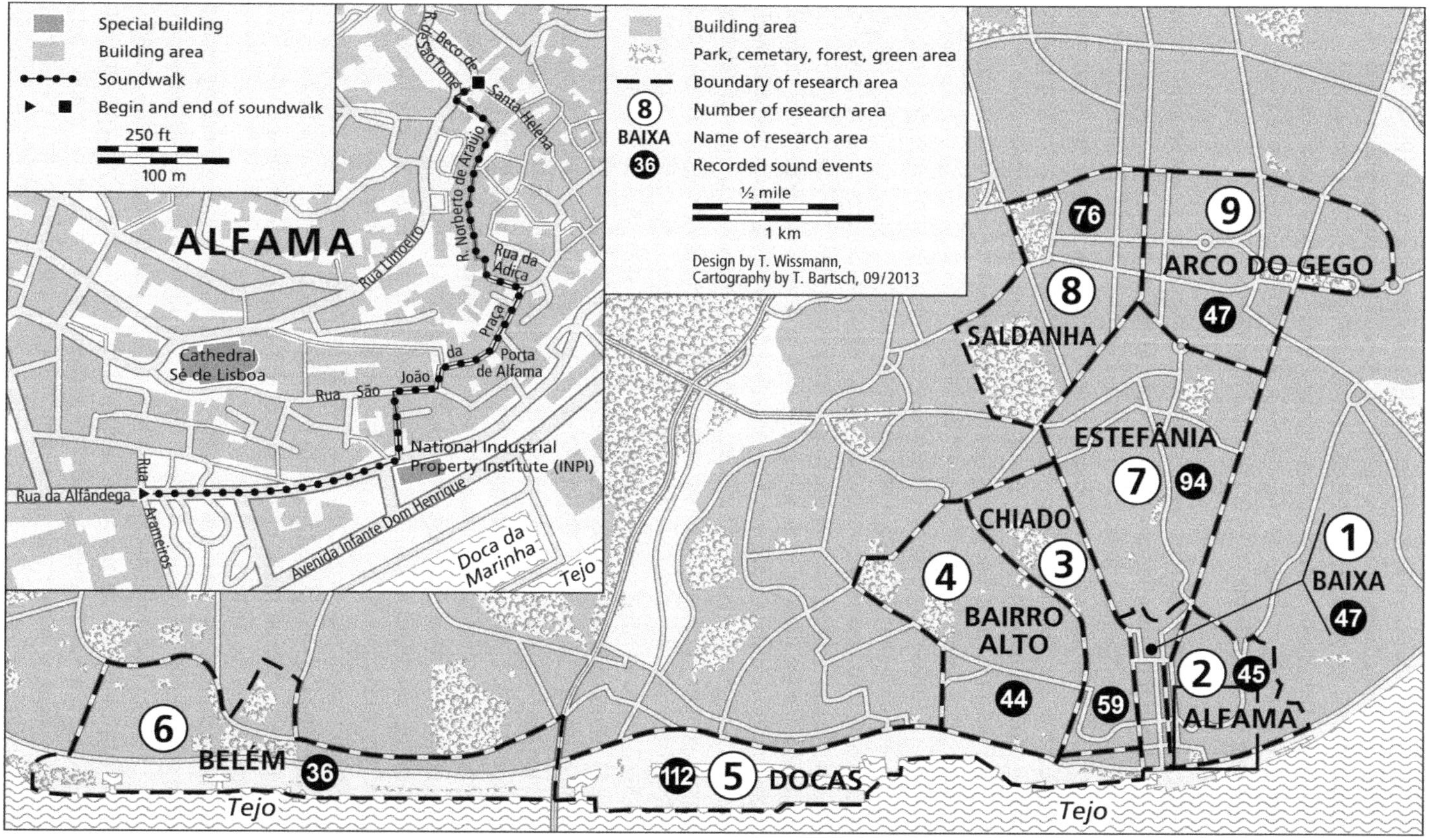

Figure 3.1 Lisbon – Soundwalk and Research Areas

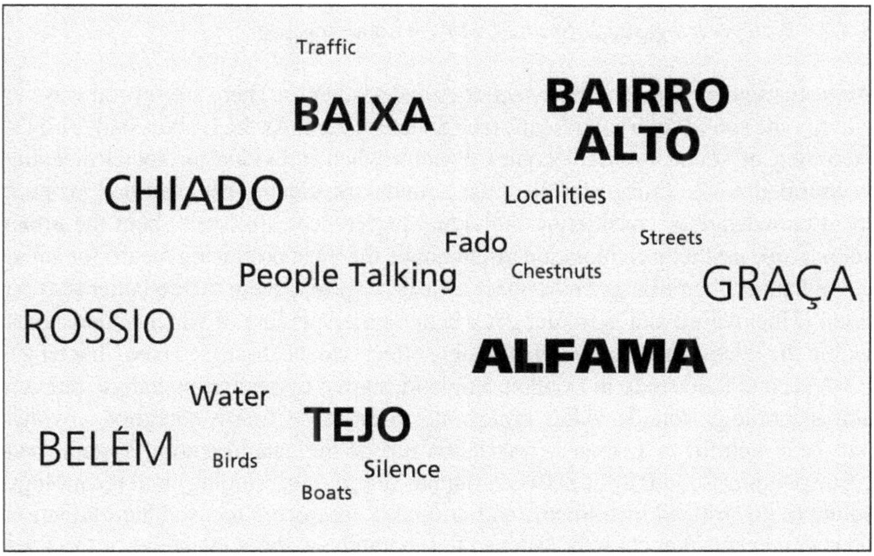

Figure 3.2 Lisbon – Sound Cloud

dominant kind of music mentioned (92 percent). Overall, Fado, the tram, and talking people were described the most significant sounds for Lisbon. The results of the questionnaire were considered for the third phase of the empirical study.

In phase three, the inner city of Lisbon was divided into nine research areas to ensure a maximum number of different sound environments for the recording (see Figure 3.2). The recording sites were not set according to a grid but rather based on the geographical appearance of relevant sound events. Over a period of three days 560 sound events were collected. Whenever multiple recordings of the site were taken, only a representative sample was documented. Recordings with poor sound quality were taken out of the study. Similar to Schafer's studies, the dates and times of the recordings were noted, as well as the type of audio recorder, place of recording, and distance from source. The track number of the stereo recordings was mentioned to link the notes back to the original sound files. Special observations and conditions during recording were also registered. Due to generalization and evaluation, 289 samples remained for further analysis. The final set of sound events gave an impression of the variety of sounds of Lisbon. Because of earlier soundwalks and the questionnaire, a representative sound profile could be gathered. The aim of the study was not to use the recordings to compose an artificial soundscape like the World Soundscape Project would have done, but to learn about sound in the urban environment and its possible deconstruction, as well as to discuss the impact of individual experiences for both respondents and researchers.

3.4.2 Categorizing Single Sound Events – London

Acoustics and psychoacoustics, semiotics and aesthetics: There are several ways to classify the sounds of a soundscape (see Schafer 1994). As the Lisbon study shows, recording of sound events becomes difficult when individual or social meaning is ascribed to it. Categories like the soundmark cannot be identified without local knowledge or considering individual preferences. To learn about the urban soundscape aesthetic connotation might not be the most promising means for a first classification. Acoustic and psychoacoustic descriptions seem to be a better starting point if the overall aim is to first get a better understanding of which sounds exist within the urban landscape before possible effects can be discussed (see Chapter 2).

A second field study in London, England, started by separating nature, human, and technology sounds. After pretesting, *music* as a fourth category – which had been helpful in Lisbon – was taken out of the categorization system, and corresponding sounds split between human sound (e.g., singing) and technology sound (e.g., stringed instruments and drums). Researchers focused their attention on sound events that could be listened to separately, without interference from the surrounding soundscape. To deconstruct the soundscape even further, the three categories were complemented with additional qualities.

- Human sound, verbal: Talking, shouting, and singing were included in this category. Sound samples like pedestrians talking and newspapermen selling journals were recorded. Also representative were announcements coming from loudspeakers in the subway and any conversation on a mobile phone.
- Human sound, non-verbal: Sound caused by the human body except any kind of speech. Rhythmic hand clapping while listening to live music and packing up produce on the farmer's market by the vendors fell into this category. Footsteps on a staircase and eating roasted almonds at the Thames river were additional examples.
- Nature sound, animate: All animals belonged to this category. A horse of the Mounted Branch, London's mounted police, was one example, as well as ducks swimming on a pond in Hyde Park, a dog barking in the streets of Wapping, and doves sitting on the street lights in Tower Hill.
- Nature sound, inanimate: Wind, water, and fire were the most typical sounds representing inanimate nature. In combination with trees, wind blowing through leaves recorded in Stamford was also counted as inanimate. A small waterfall in Hyde Park and sewage water at the London Docks were other examples.
- Technology sound: Technology sounds were not subdivided. Any mechanical, electronic, and technological source was included in this category. Recorded and live music – except singing – also fell into this category. Examples documented during field work in London were a helicopter flying over Greenwich at an estimated height of 150 m/500 feet, the beeping warning signal of a garbage truck moving backwards, music

from a CD player in front of a store in Notting Hill, and the buzz of a power socket near London Bridge.

- Two additional qualities were attributed to each sound event, regardless of whether they were nature, human, or technology sound:
-
  - Moving/not moving sound: The Lisbon study showed that many recorded sounds are caused by moving sources, and that it's valuable to differentiate between those sounds and the ones caused by objects that are not moving. A taxi driving over a speed bump near Covent Garden, a stroller rolling over the sidewalk in Garrick Street, and a seagull flying over the trees in Hyde Park were examples of moving sounds. Stationary sounds documented included a car engine starting in a parking spot in Stamford, the church bells of St. Matthew's Church in Brixton, and music coming from a restaurant at Paddington Station.
  - Singular/repetitive/permanent sound: Sound occurs in a certain acoustic pattern. It can appear as one singular event, multiple times with certain moments of silence in between, or be there permanently. The opening of a door and a car engine starting up are examples, as is a bike bell that is only rung once. During the London study, recordings of repetitive sounds included a supermarket scanner beeping and an in-use mortiser in the streets, both near Elephant & Castle. Bells ringing at Tower Hill and heels clacking on the asphalt near Paddington were other examples. Wind is hard to document as most of the time it occurs as interference to the intended sound event recording rather than as the desired event itself. If wind is recorded, it is an example of a permanent sound. Suitcase wheels on London Bridge, a ventilation system in Greenwich, and background music at a Starbuck's at Convent Garden fell into this category during the London study.

Participating researchers and student assistants discussed the application of the classification system for "arguable" sounds before field recording began. The question of source and sound caused problems when it came to assigning certain qualities to sounds. Wind, for example, is understood as an inanimate nature sound. Determining whether the sound of wind blowing leaves in a tree should be classified as "animate" or "inanimate" was not easy. In biological terms tree and leaves count as animate. For the London study it was decided, however, that they should be considered inanimate. In the context of the overall soundscape, putting that sound in the same category as wind rather than a dog bark seemed to the researchers more useful.

To cover a variety of districts, 12 research sites were set up from Stamford to Greenwich, and from Brixton to Notting Hill (Figure 3.3). On a variable route researchers documented any single sound event that caught their attention. To illustrate the sound event, additional pictures were taken and the route was mapped. All recordings were taken at a distance between half a meter (about 1.6 feet/leaves

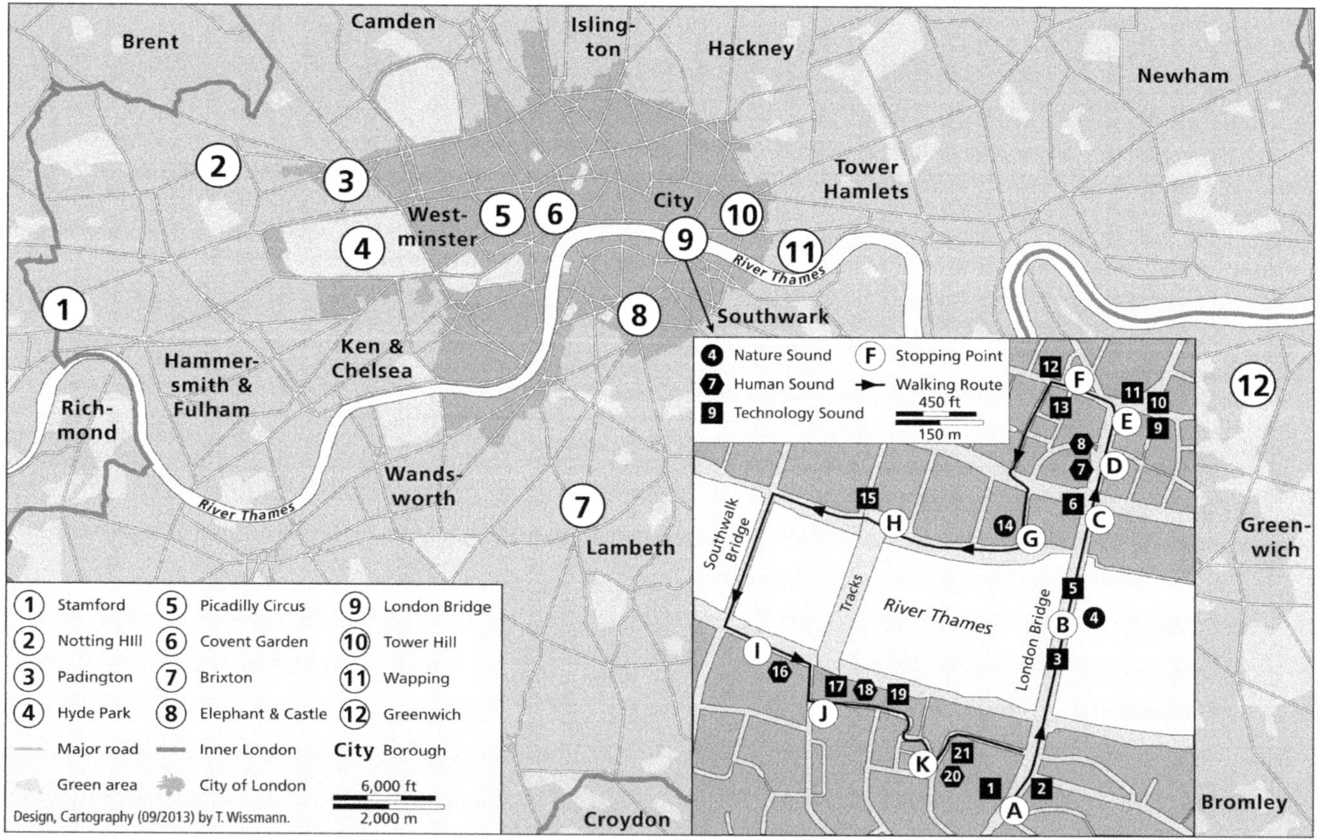

Figure 3.3 London – Research Areas, Including Area 9: London Bridge

on the ground), and 10 meters (about 32.8 feet / horn of a car). Figure 3.3 gives an overview of the recording around London Bridge.

Technology sounds occured the most, followed by human and nature sounds. Overall, 287 sound events were recorded and documented. 49 percent of the sounds fell into the category of technology, while a somewhat smaller amount were categorized as human sounds (43 percent). The numbers of verbal and non-verbal human sounds barely differed (22 percent to 21 percent of the entire collection). And perhaps a little surprising for an agglomeration like London, nature sounds formed the smallest group, with animate and inanimate sounds each making up only 4 percent of the documented sounds.

The approach to deconstructing the soundscape through the trichotomy of nature, human, and technology sounds promoted by Schafer (1994) was valuable for the ongoing study as sound recording and classification has been tested empirically. In the final discussion of the London study one fundamentally geographic question arose that could not be answered using the existing methods of soundscape recording: What are the most usual sounds at a predefined location? What are the typical sounds for an area, quarter, or district in the city?

3.4.3 Specific Sounds for Urban Areas – Austin

The Austin fieldwork builds on the outcomes of both studies in Lisbon and London. To find out more about the urban soundscape and formulate relevant research questions, five different research areas were detected (Figure 3.4).

After exploring the city in multiple soundwalks, the decision was made to record both a rather loud and a rather quiet part of Austin. Urban green space was included, as well as areas with high traffic and high population density. A typical neighborhood was selected, and Austin's emphasis on education was displayed by the researcher choosing an area with students and university buildings. As the following examples will show, the five areas each exemplified a certain part of the city and contained overlapping as well as unique sound events. The study's aim was not to describe what Austin "sounds like" because what works for artistic-like soundscape compositions in acoustic ecology won't work for soundscape research in human geography. The results would be too deterministic, and would not lead to help achieve the goal of the research project: to learn about the impact of urban sound on people living in a city.

An analysis of localized sounds is used to illustrate sound issues in certain districts of the city. Such an analysis can help sensitize ears for the local situation and raise new research questions for dealing with sound-related issues and help solve existing problems (see Chapter 4). To learn about the soundscape of an urban area, representative parts of that area have to be picked for investigation. Research areas are generalized versions of the district representing and containing all relevant soundscape characteristics. Fieldwork in research areas starts with a general inspection to verify physical dimensions and to set up a path for the soundwalk. Like in Lisbon, soundwalking is used

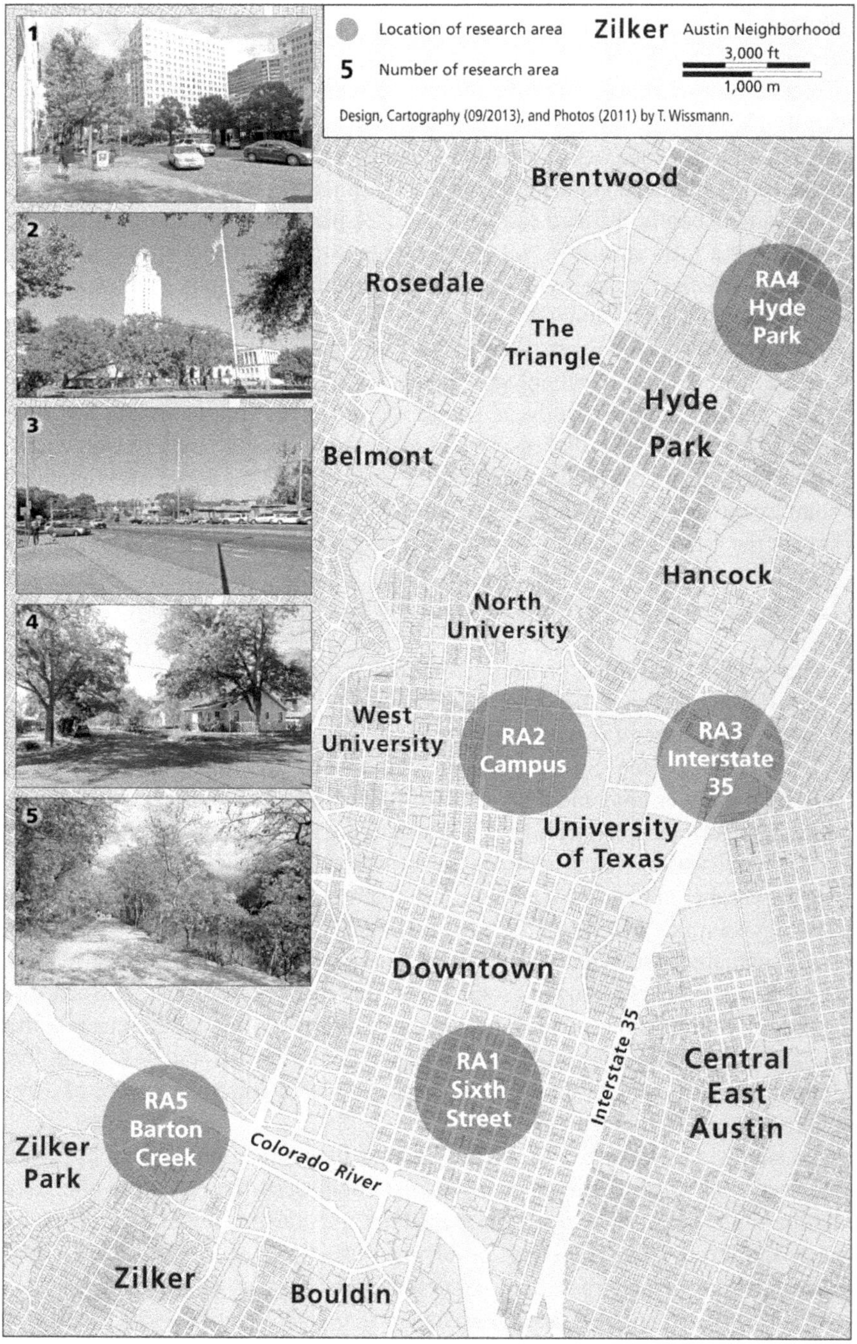

Figure 3.4 Austin, TX – Research Areas

as an initial way to learn more about the soundscape of predefined areas, but the Austin soundwalks were modified to collect additional information. In addition to documenting the walk by writing down individual impressions, the researcher stopped at multiple points to listen even closer to the sounds occurring on site. Dominant sound events – divided into the trichotomy of nature, human, and technology sounds – were geo-referenced, and additional information about the stopping points was included into a GIS on site instantly. In addition to the date, time, and position, weather conditions and individual observations were noted. A class 3 decibel measurement device was used to document the sound level (dB(A)). As opposed to measuring procedures used in acoustic design and urban planning – where a series of measurements, permanent recordings, and a time series including day, evening, night, and the seasons are used – measurements in the Austin study tended to provide an overview. The aim was not to find out about the exact noise level at a certain spot, but to establish the noise level in perspective to individual experiences of the soundscape. To complete documentation, photographs were taken at every stopping point facing the direction of the predefined route. The whole soundwalk was recorded on an audio device.

After audio-visual documenting was done, the data was evaluated to determine the most dominant and typical sounds in the area. Mapping, as well as photographs, audio recording, sound level measurements, and the protocol of the individual experience were used to locate one single point on the route that was representative for the area, and thus for the whole urban district. Findings made possible the next step of the fieldwork: *sound point recording* (see Figure 3.5). A *sound point* is a position within a predefined area where the typical soundscape of that area can be experienced. It may be located at one of the stopping points on a soundwalk or at an individual place that combines most of the collected data in a representative way. To record the soundscape at the sound point, a microphone was fixed on a tripod for the Austin study. As the different research areas had to be comparable, the recording time was identical (midday). The duration of the recordings was one hour. The results differed in various aspects, as the following descriptions show.

Research area 1: Sixth Street

Sixth Street is part of downtown Austin, covering roughly 10 × 10 blocks of the capital of Texas. Between the Colorado River and First Street in the South and the *Texas State Capitol* at 11th Street in the North, between Nueces Street in the East and Red River Street in the West, the core of Austin is located. It contains historic buildings, is an area for business and administration, and has two nightlife districts: the Warehouse District and Sixth Street. Many streets running north/south experience heavy traffic – like Guadalupe Street, Lavaca Street, and Congress Avenue – while east-west streets are generally quieter. The area around Sixth Street was picked by the researcher because it contains both very loud sections, with heavy traffic, pedestrians, and music, and also more quiet sites where nature sounds dominate the soundscape. As Figure 3.6 shows, research area 1 covered three blocks. The soundwalk started on Brazos Street and made its way from Sixth

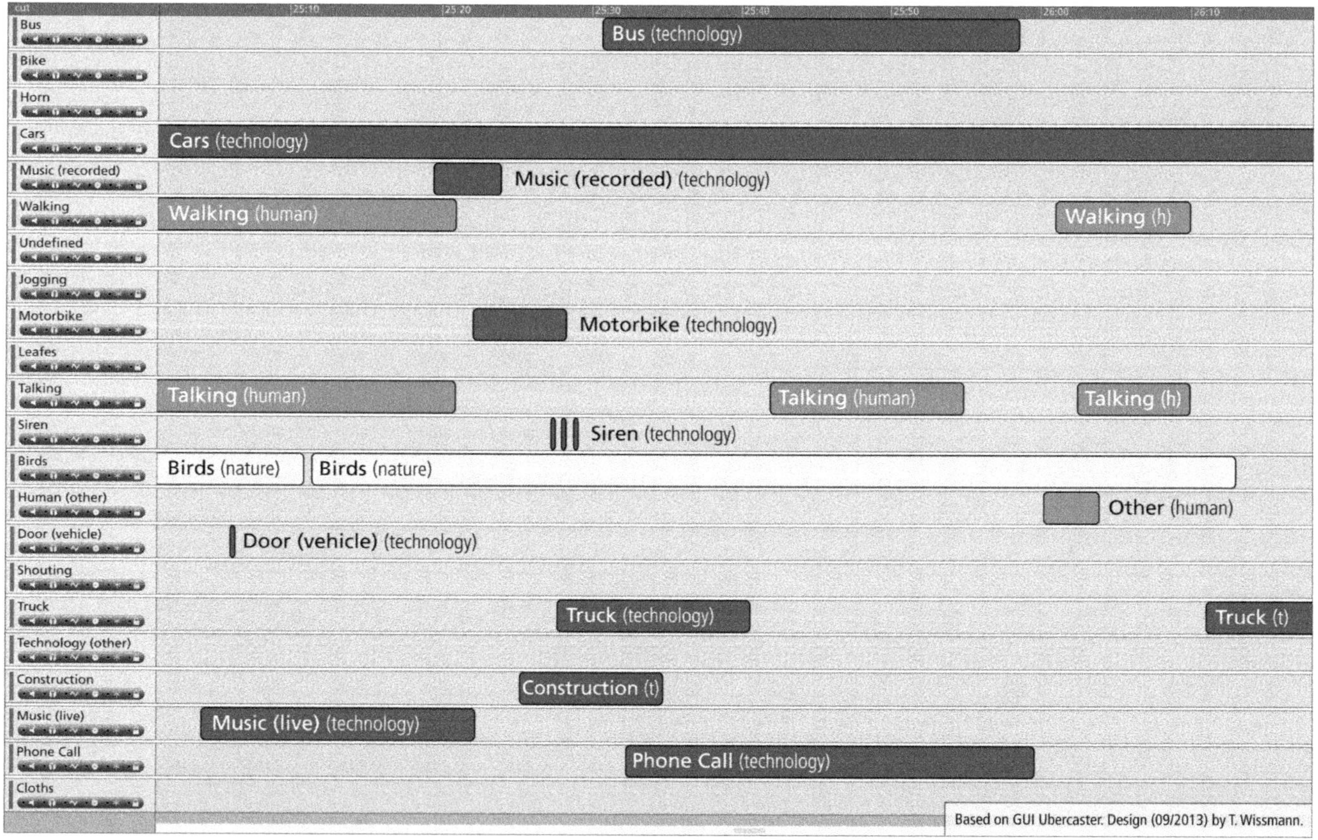

Figure 3.5 **Sound Point Recording – Sequence From Research Area 1**

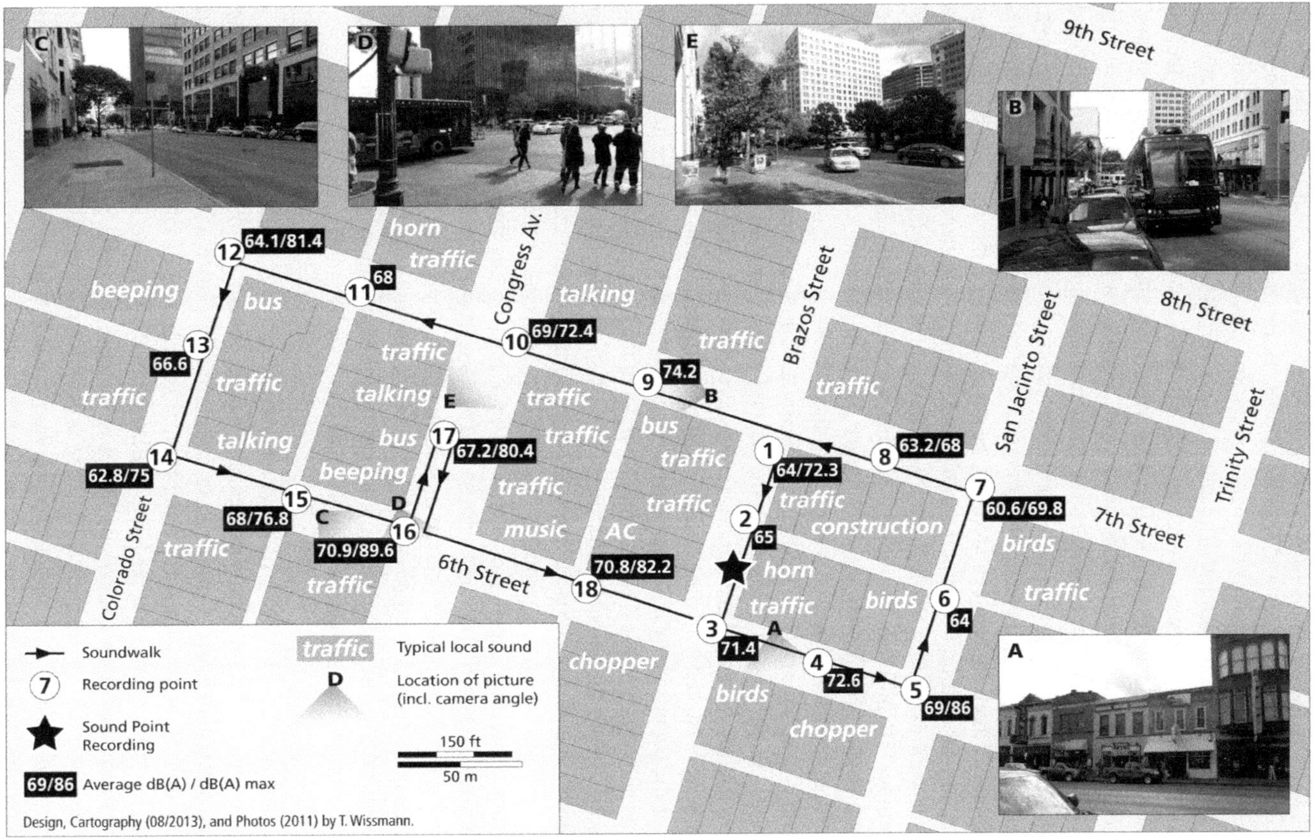

Figure 3.6 Austin, TX – Research Area 1: Sixth Street

Street, San Jacinto Boulevard, and Seventh Street to Congress Avenue, where it continued west on Seventh Street to Colorado Street and then Sixth again. After a short detour entering Congress Avenue from the South, the soundwalk continued on Sixth Street and finally stopped at Brazos Street, where recording started. At 18 stopping points the local soundscape was documented, at a distance of half a block from one another. The sound levels varied from almost 90 dB(A) near a bus station on Congress Avenue to only 60 dB(A) at the corner of San Jacinto Boulevard and Seventh Street. The area was dominated by technology sounds caused by cars, busses, and construction work. People walking the streets and having conversations with other pedestrians or on their mobile phones were common sound events. Birds almost exclusively represented nature sounds in the downtown area. In the analysis of sound events and their occurrence, the sound point was identified on Brazos Street, in between stopping points 2 and 3 of the soundwalk. The bus route, constant traffic flow, and hearable birds were strong factors for choosing the site, as well as nearby Sixth Street, with its high number of pedestrians. During recordings 10 different technology sounds were documented, followed by four human sounds. Birds were the only nature sound in the area. Cars were by far the most dominant sound source, with appearances 97.94 percent of the time. Birds, in second place, occurred 68.78 percent of the time, followed by talking people (26.26 percent). Even though birds were the second most common sound in the area, nature sounds were the least important in general. Overall, only 22.92 percent of the sounds were caused by nature, while 25.20 percent had a human source. More than half of the soundscape (51.87 percent) was shaped by technology. The high number of cars and other vehicles, responsible for most of the high sound level results, raises questions about the impact of traffic on the urban dweller. This seems to back up the theory that noise-exposure maps and regulations to minimize sounds caused by traffic are an issue in city planning. This will be discussed in the next chapter (4.2). Another interesting question concerns the branding of a city through sound. As the soundwalk followed a part of Austin's most famous live-music district, Sixth Street, the marginal appearance of music in recordings (4.47 percent) could be explained as a result of the time of the recording (midday). But for a city that labels itself the Live Music Capital of the World, a tourist might expect more when visiting Sixth Street even during the day. Section 4.3 will focus on live music.

Research area 2: University of Texas campus

The campus of the University of Texas at Austin contained not one but two of the selected research areas. Compared to area three, Interstate 35, which focuses on traffic sounds, area two captures the typical soundscape of campus life. With an extension from Martin Luther King Boulevard in the south to Dean Keeton Street in the north, and from Guadalupe Street in the east to Leona Street in the west – on the other side of Interstate 35 – the university campus all but encompasses the core of Austin. It is surrounded by major roads with a lot of traffic sounds and contains pedestrian areas and parts where cars with a special permission and city busses are allowed to drive. Research area two (see Figure 3.7) covers all

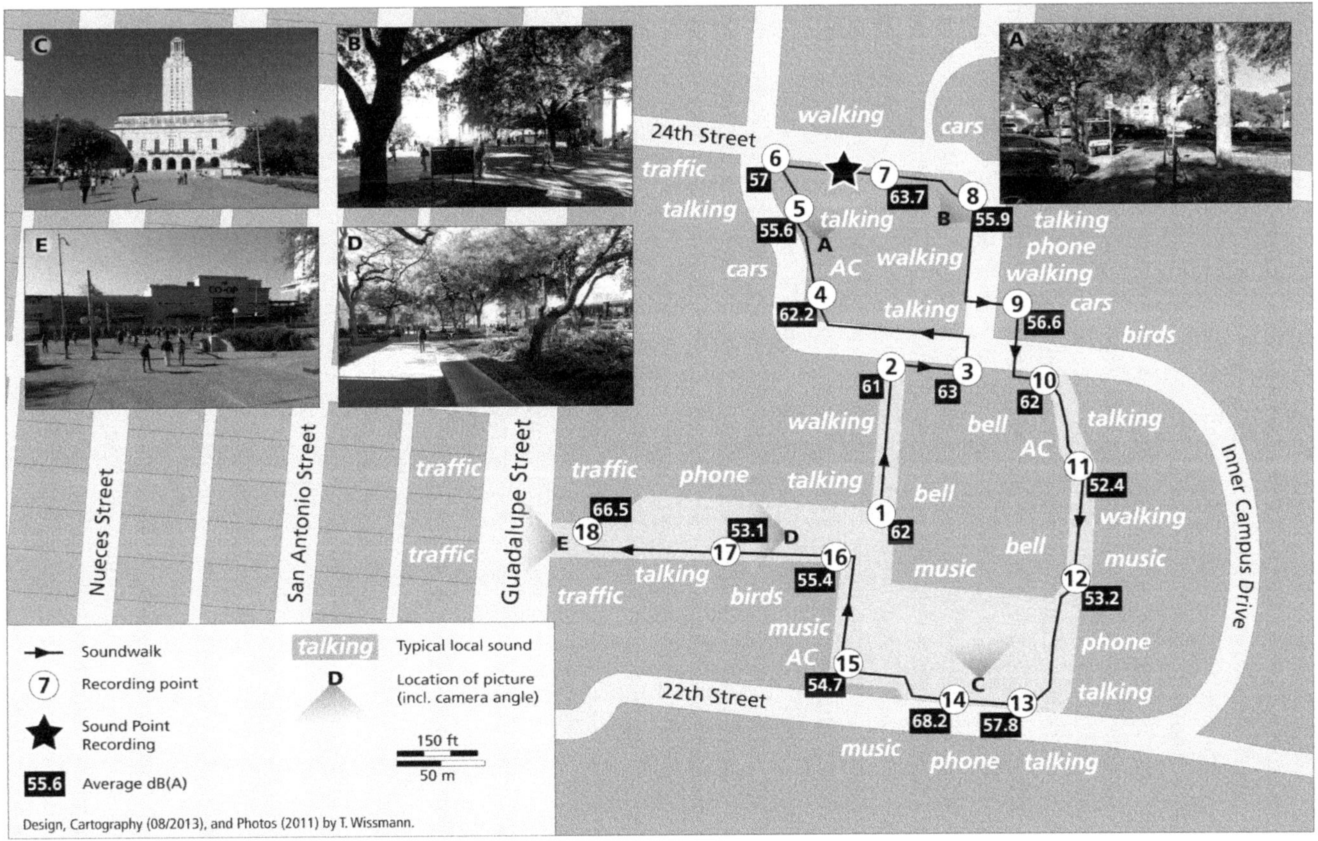

Figure 3.7 Austin, TX – Research Area 2: University of Texas Campus

these different soundscape elements. At the western border, Guadalupe Street is dominated by technology sounds, while on the east side – close to the university tower – mostly human sounds are heard. Nature sounds occur all around the area and can especially be spotted near walkways where no cars are allowed. From 24th Street, were vehicles can enter the campus and the Department of Geography and the Environment is located, the area extends southward to 22nd Street.

The soundwalk started at the west side of the university tower and followed Inner Campus Drive to 24th Street. The route turned right and turned back to Inner Campus Drive at Mary E. Gearing Hall. Upstairs and over a small pathway the route continued around the university tower to the pedestrian area that leads directly to Guadalupe Street, where the soundwalk ended. The loudest sound was recorded on the plaza in front of the tower, where students played music over multiple loudspeakers (68.2 dB(A)). The pathway east of the tower was the most quiet area (52.4 dB(A)). Human sounds were most dominant, with people walking, talking to each other, and having cell phone conversations. Also common were the sounds of birds and leaves in the trees, but also cars close to Guadalupe and running air conditioning systems all over the defined area. For sound point recording a site on 24th Street was chosen, as all relevant sounds could be found there. The sound of people walking was by far the most dominant one in the research area (69.83 percent). Bird sounds followed in second place with 55 percent, and talking covered 26.26 percent of the sound point recording's time. The presence of two human sounds within the top three indicates the overall dominance of this category. Although 10 different technology sounds were recorded in contrast to only five human sounds, the latter's 48.12 percent of the soundscape more than doubled technology's 22.21 percent. Even nature sounds occurred more often, 29.17 percent. A look at the three most dominant sounds in research area two not only indicates the dominance of human sounds, but combined with visual observations during sound point recording also raises an interesting question. While almost 70 percent of the time walking people were recorded, only around 30 percent of the time did talking occur. Given the fact that not only moving (i.e., walking) people talk, but also students sitting along campus roads and pathways, many of the walkers were silent. Adding cell phone conversations (about 9 percent) still left a high number of not-talking but walking people, many of those wearing earphones. It cannot be said whether audio content was being played back over those earphones at those times, but it can be assumed that most students wear those devices only while listening to audio content of some kind. The creation of an individual soundscape has already been described in section 2.5, and Chapter 5 will take on the issue with a media geographic discussion on sound. Aside from listening to podcasts and music, earphones can also be used to listen to audio-guided tours, like the *Audio Tour for Works of Public Art* produced by Landmarks (n.d.), that complements public art installations on the university campus. Chapter 4, section 4.4 will discuss the issue of audio guides in detail.

Research area 3: Interstate 35

The second area on campus was chosen because it contained the sounds caused by traffic on major roads, highways, and interstates. Thus, Interstate 35 is less cubic than the four other research areas. It follows Dean Keeton Street, covers a part of the parking area of the Lyndon B. Johnson Library, and follows Interstate 35 for a while (Figure 3.8). Starting at the corner of East Campus Drive / Medical Arts Street and Dean Keeton Street, the soundwalk continued down the latter to Red River Street. Turning right it entered the parking area and circled back to Dean Keeton. Over a walkway the route reached Interstate 35. Recordings were made along the highway. Back on Dean Keeton the soundwalk continued under the Interstate to reach its end on the east site of 35. The most quiet location was recorded at the south-west corner of the parking area (58.2 dB(A)). Sound level measuring at about three feet (i.e., about one meter) from an Interstate lane showed 93.7 dB(A), the overall maximum value for all five research areas. Traffic sounds were so dominant that additional subcategories were created to better capture the different sound impressions. Traffic from Interstate 35 was measured separately from the cars on the other streets. Traffic waiting at lights produced sounds that were also excluded from the general *traffic* category. For sound point recording a site away from the interstate was selected, to include but not overrate the latter relating to the overall sound situation in Austin. Analysis shows that 11 different technology sounds added to the soundscape, combined with four human and three nature sounds. Nature and human sounds together accounted for about 14 percent of the soundscape (4.91 percent and 8.65 percent, respectively), while 85.96 percent were technology sounds. According to this, the three sounds occurring the most were Interstate 35 traffic, which is heard all the time (100 percent), cars on other major and minor roads (69.64 percent), and traffic waiting at lights (42.92 percent).

The results emphasize the need for a discussion about traffic and traffic noise. As section 4.2 will show, traffic plays a major role in discussions about noise in urban environments. It is essential to urban planning and can cause serious hearing loss and other physical trauma, as section 2.2 has shown.

Research area 4: Hyde Park

While traffic sounds along Interstate 35 produced the loudest sound samples of all recordings in all research areas, Hyde Park contained the quietest ones. Hyde Park is an old neighborhood north of Central Austin and the university campus that consists of mostly single-family homes and a lot of vegetation like trees, bushes, and grass. Its southern border is 38th Street, from which the neighborhood extends northward to 51st Street (Figure 3.9). Guadalupe Street is the western border, while Duval Street and Red River Street define the area in the east. The researcher selected the two blocks between 48th and 49th Street, Duval Street and Avenue G, because they are the quietest parts of the Hyde Park neighborhood. The bus lane on Duval Street and the traffic stands 51st Street are the loudest areas of the district. The first stopping point of the soundwalk was the corner of

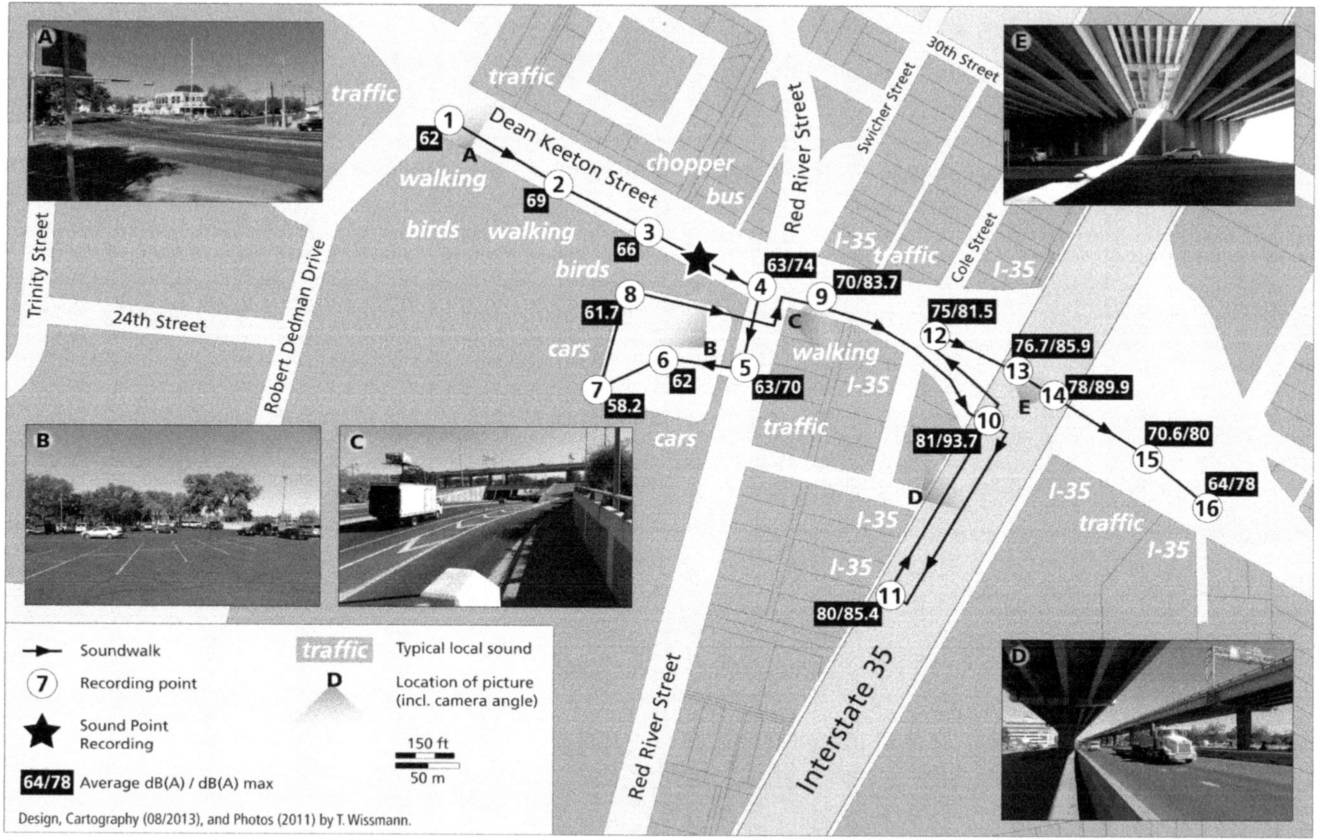

Figure 3.8 Austin, TX – Research Area 3: Interstate 35

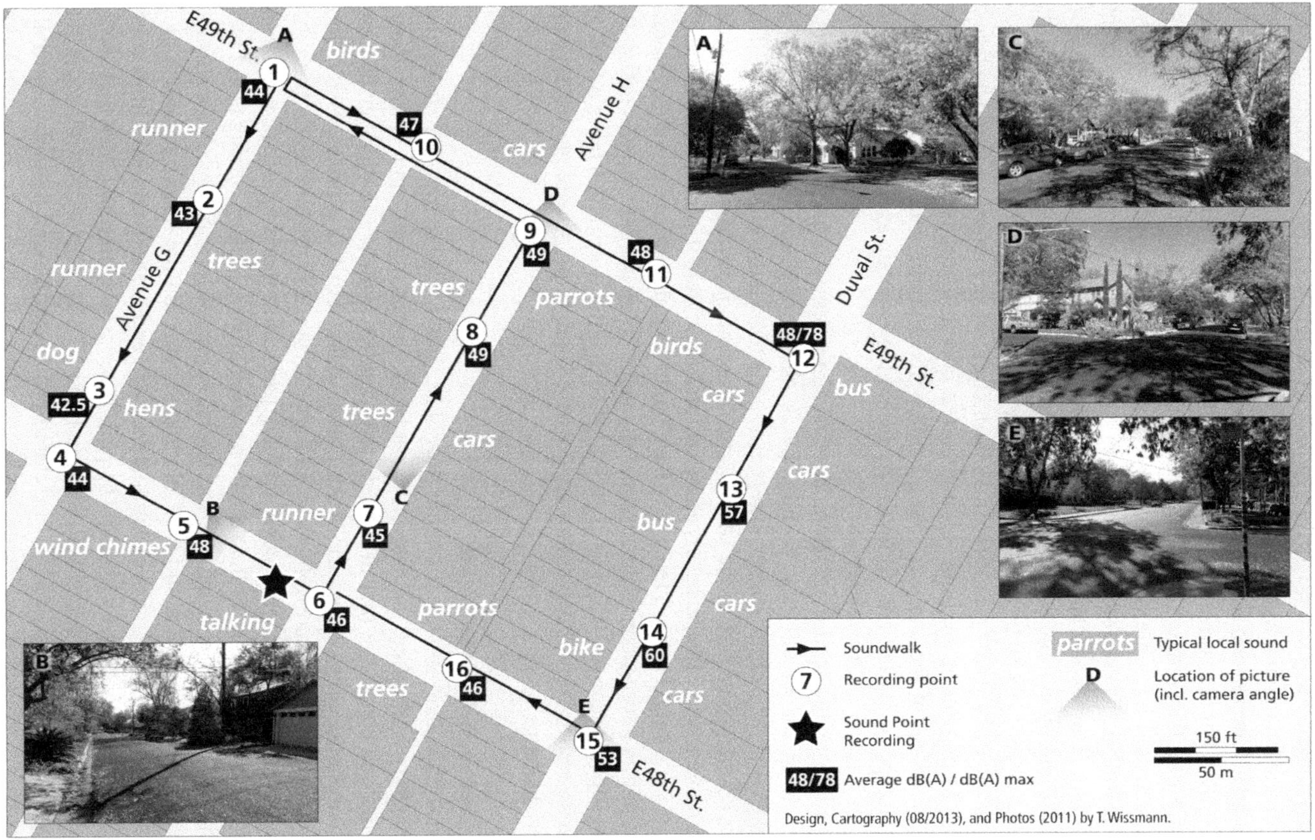

Figure 3.9 Austin, TX – Research Area 4: Hyde Park

Avenue G and 49th Street. Down the avenue, two stops were set up, each about one-third of a block in length. At 48th Street the route went left, entered Avenue H, and continued up to 49th Street again. After a right turn the soundwalk reached Duval and followed it to the corner of 48th. At the crossing of Avenue H the route ended. Different sound levels were recorded, starting at 42.5 dB(A) on Avenue G to 78 dB(A) at the corner of Duval and 49th Street (48 dB(A) in periods without any traffic). Birds, dogs, and trees were typical nature sounds found in the area, supplemented by some parrots and evens hens. From the data collected on the soundwalk, a site on 48th Street close to Avenue H was selected for sound point recording. The soundscape was mostly quiet there with occasional appearances of cars driving by and the sound of busses from Duval. Almost all of the time birds could be heard (99.86 percent), followed by cars (17.81 percent) and the sound or leaves and trees (14.06 percent). As a seasonal sound, lawnmowers could be heard about 33 percent of the time. As it occurred only very occasionally, this source was taken out of the analysis. Thus, 68.48 percent nature sounds dominated the soundscape of Hyde Park, followed by technology sounds at 19.97 percent, and human sounds, 10.20 percent.

Unlike in the recordings from the first three research areas, one important element of the Hyde Park soundscape is silence. Silence, which cannot be displayed by any of the chosen categories, adds to soundscapes nonetheless. The focus on silence as part of the soundscape raises the question: Where in an urban environment can silence be experienced, and do different qualities of it exist? Silence in a neighborhood like Hyde Park might include soft sounds of leaves and wind and is never a complete absence of sound. To investigate the perception of silence and the impact it can have on the urban dweller, section 4.5 takes a look at churches in the Austin downtown area, where it might be possible to experience "sacred silence".

Research area 5: Barton Creek

To integrate one of Austin's multiple public green spaces into the study, the researcher established the last study area in Barton Creek and Zilker Park. Springs and creeks are part of the Edwards Aquifer, which, among other factors, geomorphologically shapes the Austin area. Zilker Park is a prime destination for joggers, cyclists, and walkers to find quiet, relax, and refresh. Aside from the open green there are a lot of trees, walking paths, and sports fields. The research area was picked because of the various representative sounds occurring there, like traffic from Barton Springs Road, the MoPac expressway, and Downtown Austin; water from Barton Creek and the Colorado River; multiple leisure localities like baseball fields and the tracks of a miniature train; and the hike-and-bike trails used by joggers, cyclists, and walkers. The park is set around Barton Creek and starts at a baseball field on Toomey Road (see Figure 3.10). The soundwalk route set up by the researcher turned left to Colorado River and passed two other baseball fields before joining a small path leading to the Lady Bird Lake Hike and Bike Trail. The soundwalk followed the trail to the mouth of Barton Creek and left

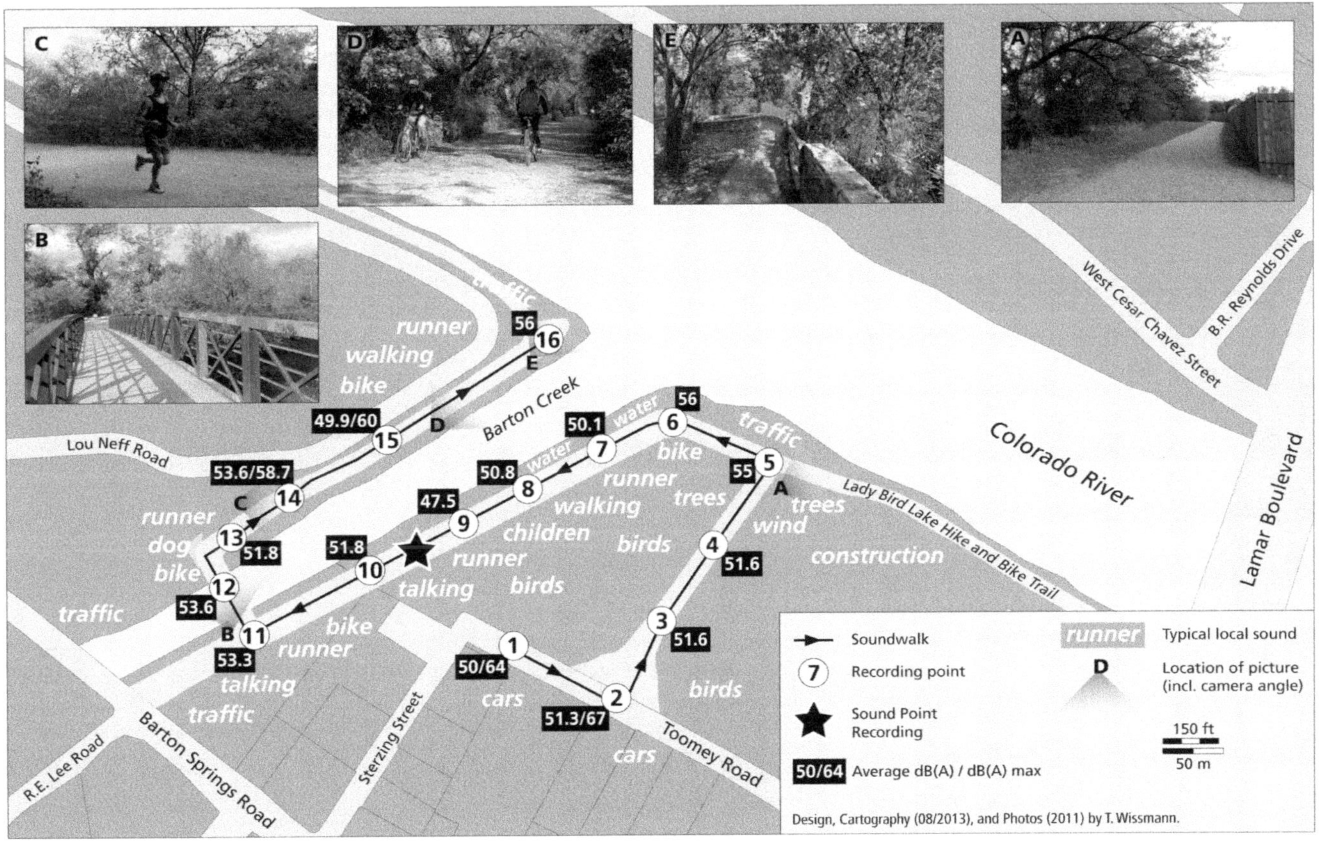

Figure 3.10 Austin, TX – Research Area 5: Barton Creek

the Colorado River to follow the creek upstream. Before the trail crossed Barton Springs Road the route turned right to cross Barton Creek over a footbridge. On the other side, the soundwalk continued downstream between the creek and one of Zilker Park's major open areas. It then ran alongside the tracks of the Zilker Zephyr Miniature Train. The soundwalk ended at Lou Neff Point on the northwest side of the creek's mouth. The loudest site on the soundwalk was the ending point at Lou Neff Point (56 dB(A)). The quietest site was on the northwest side of the creek (49.9 dB(A)). The research area was a homogeneous one in terms of sound level, with a relatively small variation of volume (only around six decibels) among the 16 measurements sites of the soundwalk. Joggers, walkers, and cyclists could be heard at almost every spot on of the soundwalk, while the area around Toomey Road and Barton Springs Road was partly dominated by the technology sounds of traffic. Sound point recording was conducted in the middle of the southeast bank of Barton Creek, on the hike-and-bike trail. Bird sounds occurred the most in recordings (37.36 percent), followed by the two human sounds of walking (15.44 percent), and jogging (14.69 percent). With talking in fourth place, human sounds constituted the majority of the soundscape (36.88 percent), closely followed by nature sounds (36.12 percent). Technology sounds accounted for 23.05 percent of the soundscape at Barton Creek. The relative balance of sounds also showed in the different sounds recorded, with four nature, five human, and seven technology sounds. In all other research areas – even in Hyde Park – technology sounds exceeded all human and nature sounds combined, and sometimes even more than doubled them.

Against the rather quiet soundscape of Barton Creek, temporary sound events stuck out that could not adequately be displayed by percent of the soundscape. 1.08 percent of the time a plane or helicopter could be heard. Sounds of airplanes flying over an area can have a huge effect on the listener and impact the soundscape significantly due to individual perception and subjective connotation. A field study in the Rhine-Main-Area in Germany investigates the issue, using the example of air traffic sound caused by Frankfurt Airport (section 4.2.3).

All research areas represent a larger part of Austin and possess an individual and characteristic soundscape. With sound point recording it is possible to document the typical soundscape of an area (see Figure 3.5). Thus, each sound point produces relevant data for the soundscape of the entire city. Still, sound point recording can never display how a city sounds, i.e., what a typical "sound of Austin" would be. Sound point recording should rather be used to generate research questions based on a new sensitization regarding urban sound. Based on the five Austin research areas, the following issues stand out:

- How important is a city brand like Austin's "Live Music Capitol of the World" for the local population?
- How does sound in media affect the perception of everyday life?
- What impact do traffic sounds have on the organization of a city and the quality of life of its inhabitants?

- What part does silence play in the urban soundscape, and what different qualities can it have (so that certain types, like sacred silence, might be experienced)?
- How does air traffic sound affect the residents living in the entry lane of global airports?

These and more questions will be discussed in the next chapter, which will focus on the various uses of sound in the city.

Chapter 4
Uses of Sound in the City

4.1 Managing Sound in the Urban Environment

There are not only various sound sources that build the urban soundscape, but also numerous ways these sources can affect the urban dweller. As described in section 1.4, many scientific disciplines are dealing with the topic, from acoustic ecology, urban planning, and architecture to acoustic design, informatics, musicology, ethnology sociology, and psychology. The always-different impact of sound has been pointed out, and unwanted sounds have been contrasted with the liked or wanted ones. As part of the city, sound is always present and can be deconstructed into the three categories of technology, human, and nature sounds. Generally, sounds belonging to the first two categories dominate the urban soundscape. Only in more quiet environments like parks, recreation areas, and certain neighborhoods, can the urban dweller experience nature sounds in particular. Most of the time, the singing of birds and the gurgling water of streams and creeks – if present at all – are masked by the higher sound level of vehicles, construction areas, and general urban hustle.

How does the city and how does the urban dweller deal with this cacophonic mix? If the environmental soundscape is not replaced by an individual playlist coming from the earphones of a Walkman and iPod, urban sound influences how the city is being perceived and directly affects the listener. So, how can developers and governments manage the urban soundscape in a way that allows more and more people to enjoy it and not have to suffer from secondary effects like sleeplessness, bad temper, or the impossibility of using outdoor areas like their own front gardens? Noise may be the most obvious sound that has to be taken care of. City planning often has only few options for creating a pleasant soundscape. Due to the general layout of a city, its topography, roads, and buildings, the only possible way to enhance the soundscape might be to add additional noise insulation. Along inner-city highways and other highly frequented roads, walls can be put up to mask most of the noise. Older buildings in particular do not contain proper noise insulation. As Chapter 4, section 4.2 will reveal, many newly constructed buildings in the United States also lack insulation. If condos and one-family homes are supposed to generate a maximum amount of profit for the building enterprise, it is no wonder that such features are not included automatically. An apartment that looks and sounds nice during the day might be perceived rather differently when live music from multiple venues is playing every night.

Urban green is another way to enhance the sound quality of a city. Works on the "effects of vegetation on the perception of traffic noise" (Watts, Chinn, and Godfrey 1999: 39) state that additional trees and bushes reduce the measured sound level coming from streets. But they also conclude that:

> narrow belts of trees and shrubs should not be used for noise screening purposes in the belief that although they provide little physical noise reduction they improve the perception of the noise through the mediating effect of improved visual appearance (Watts, Chinn, and Godfrey 1999: 55).

This example again shows the complexity of managing sound in the urban environment. While the sound level is reduced, individual awareness might be raised. A sound that is not seen has to be listened to more closely than one whose source can be identified visually.

The behavior of park visitors has changed over the years. Today, less people use urban green spaces to relax in silence (see Institute for Environment and Health 1997: 5). Their voices add to the soundscape, either in face-to-face discussions, phone calls, or singing. The latter is an issue wherever live music is part of the urban culture. In cities like Austin – and this will be discussed in section 4.3 – every night about 200 live performances take place. Areas dense with music venues produce a mix of multiple music genres, styles, and tempi. While both traffic and natural soundscapes are quieter at night, the beats coming from huge boxes and amplifiers appear more dominant than they would during the day. Finding a compromise between the expression of musical culture and an individual's night's sleep is a challenge for the city government. To control the general sound level at different times of the day, allowed levels of sound exposure can be measured and displayed on thematic city maps. What is a common practice within the European Union, where bigger cities and agglomerations are asked to make available such maps for the public, is missing in most cities in the United States. Even Austin, which brands itself as the Live Music Capital of the World, is missing a noise exposure map. Exceptions are made for airports, which produce high sound levels. A study in Frankfurt am Main will discuss the issue intensively, focusing on the (negative) effects of air traffic noise in the Rhine-Main Area, where Frankfurt Airport – one of the biggest passenger and cargo airports in Europe – is located.

Noise exposure maps are one way to make the topic of traffic noise transparent. Those these maps generally include data from street, air, and rail traffic, human and nature sounds are generally not part of these maps. And if there is a permit required to run cars faster than a certain decibel level at a certain time of the day, those restrictions do not relate to sources like live music. Sound on private property is also an issue. While the physical measurements of an estate are easy to name, the individually perceived impact of sound coming from a front porch or backyard is more difficult to communicate. If the city ordinance defines the maximum sound level for private properties, the implementation of such a rule is not as easy. Complaints by neighbors have to be checked, which means manpower, time, and knowledge about sound will be needed. The policeman who is called to control the compliance of noise regulations on private properties has to know how to measure sound and what to make of a numeric value on the logarithmic decibel scale. What kind of strategies a city can develop to cope with this issue will be discussed on the following pages (Chapter 4, sub-section 4.2.2).

An interesting way to make use of sound in the city is to market it. Authenticity and sense of place are both created and represented in the 'typical' music of cities like New Orleans and Detroit, Seattle and Rio, Berlin and Lisbon. Fado, a bright-melancholic music from Portugal's capital, acts as a soundmark for the whole city. Visitors coming the Lisbon want to see certain landmarks like Torre de Belem – the place where Christopher Columbus started his famous voyage – but also listen to typical Fado music. For about 20 years, Austin has branded itself the Live Music Capital of the World. The Austin Convention and Visitors Bureau (AVCB) recognized that the acoustic guitar – a symbol for Americana and country music – was a powerful tool to attract visitors nationwide, and even worldwide. Several campaigns promote the city through live music, supported by the large number of live music venues and internationally recognized music festivals like Austin City Limits (ACL) and South By Southwest (SXSW). Music is used as a marketing tool to boost tourism but also create an environment for artists and young professionals. Labeling a city in such a way makes it look open minded, young, and creative. Musicians and other artists are attracted to such an environment, as are certain demographics that seek the creative potential of a city.

How locals and city officials make use of the urban soundscape determines how this soundscape is being perceived. Chapter 2, section 2.5, showed how easy it is nowadays to tune out unwanted environmental sounds by putting on earphones (Bull 2007). If sound sources like live music venues are positively labeled, the sound coming from guitars, drums, and basses are less likely to be perceived as noise but rather wanted sound. If the local population consists of a large number of audiophiles, complaints about nighttime disturbance of the peace through music might be less of an issue. And if the musicians playing in the music venues become famous, they will not only be associated with their hometown (of choice) but also have a positive marketing effect on the city itself. Every musician touring nationally or internationally implicitly acts as an ambassador of a city's brand.

Tourists arriving in a city consume not only music but can use the sound of audio-guided city tours to get familiar with the surroundings and typical local elements like landmarks. Completely on their own, only using a prerecorded audio file and map for directions, visitors rely on sound to explore the urban environment. The technology, which was pioneered in the Sedelijk Museum in Amsterdam in 1952 (Tallon), offers touristic information service free from restrictions related to day and time. The narrator's voice replaces the tour guide as a social actor, speaking directly to the listener. The propagated sense of place comes from a script that presents the city in a way the producer wants to market it. The power inherent in the audio-guided tour will be deconstructed in section 4.4. The possibilities of using audio-guided tours both in museums and outside as an asset for city marketing will also be investigated.

One thing an audio-guided tour cannot replicate is the silence of certain places. Tranquil space has already been discussed and quiet spaces are important for the urban dweller to relax (see Campaign to Protect Rural England 2005: 3). As absolute silence can never be created in an urban environment – aside from in laboratory experiments – public space is rather loud. More quiet places can be found inside some buildings. In libraries visitors are supposed to speak with a gentle voice to not disturb other people. Sound is used to provide a certain kind of privacy to study, think, and concentrate. Political buildings are quiet places, too. It is a form of silence that is used to demonstrate political power when visitors to a state capitol are asked to keep their voices down. The importance of a building is increased by the acoustic demonstration of power, imposing modifications on visitors' general communication behavior. The mandatory silence of libraries and political buildings can also be found in sacred places. In churches, for example, silence plays an important role. The religious experience of a house of God usually is one of silence. Praying in many cultures is a non-acoustic act. Section 4.5 will examine the sounds of sacred silence in a dual way. The recorded soundscape and experience of a church visit will be described, while the silence within the sounds of the service is also a subject of the inquiry.

Church bells are an example of the use of sound in the expression of power in the urban environment. Wherever particular bells are heard the listener realizes that s/he is in a Christian area. The proclamations of a muezzin have a similar meaning in another cultural context. Both sounds are signals for a – at least acoustic – supremacy and represent an active occupation of (urban) space. Other bells, shouts, or sirens are signals that describe specific actions and events. In times of war sirens are heavily used to warn the general public against air strikes. The siren on an ambulance is used to clear the way between the site of an accident and a nearby hospital. A similar meaning is inherent to the sirens of fire trucks and police cars. Sound signals are generally a way to organize society. They are an expression of our dos and don‘ts, e.g., the acoustic signal of a traffic light that gives pedestrians permission to cross the street. There are curbs on the roads that in every sense help us to "stay in line".

The temporary rise of sound levels due to acoustic signals describes an urban soundscape that is never static but always in flow. The dynamics can be experienced over time, especially in the course of a day. During the day the acoustic environment is far louder than at night, with several peaks like rush hour. The experience of single sound events also varies depending on the time of day. The change of the seasons is another factor co-determining the perception of sounds. Snow, for example, reduces the maximum sound level, while rain alters the sound coming from sources like driving cars and masks more quiet sounds completely. Seasonal sound is not manageable by the city council, but its daily appearance is. There are noise regulations based on the time of day and maximum decibel level, differing throughout the city. Neighborhoods with single-family homes have to be quieter than a lively downtown area, or an inner-city highway. Traffic has to follow those restrictions, leading to regulations like the prohibition of aircraft grounding during the night. Special noise protection areas are created to limit the maximum disturbance of residents (see sub-section 4.2.3).

Another drastic change of the urban soundscape takes place during cultural happenings like festivals and parades. An unusually high number of people are in the streets, and sometimes the visitors even outnumber the local population. Major events like the Austin-based Austin City Limits (ACL) and South By Southwest (SXSW) festivals attract hundreds of thousands of people every year, most of them coming from out of town. During several days around 1,000 bands perform on over 50 stages. The sound coming from the music venues exceeds the usual sound level of the area, so the city council enacted a special noise permit effective only during the few days of the festivals.

How sound can be used to influence the noise emitted from traffic can be seen in publications like the "Validation of a complex urban noise model close to a road" (Walerian et al. 2011) or "The effects of vegetation on the perception of traffic noise" (Watts, Chinn, and Godfrey 1999). Indeed, the city has a lot of options to influence and shape the sound emerging from traffic. Aside from the above-mentioned noise protection walls and special insulation of buildings, paving is a strong factor in the regulation of noise levels. Whispering asphalt emits – as its name indicates – relatively little sound, while the cobble stone pavement in European cities like Paris boosts the acoustic impact of cars.

The soundscape of stores, shops, and malls is another element where urban sound is actively used. Store music being used to influence shopping behavior and to enhance the buying experience has already been discussed (see 2.3.1) (Turley and Milliman 2000). But music is not the only element added to the soundscape. Language becomes important wherever people want to buy and sell goods. In a country like the United States social status is hearable in malls. Stores like Macy's have all their information written in English. There are no Spanish announcements over the loudspeakers. The experience in a store like Sears is different. In cities like Austin, the company puts up signs in both English and Spanish. Salespersons and announcements orient themselves to the high number of customers of Latin American heritage. Supermarkets like Fiesta Mart focus on the Hispanic population and not only sell goods from Latin American countries but also organize their stores with Spanish signs and employ a high number of Hispanic salespeople. Due to this use of the Spanish language the store's soundscape significantly differs from other food supermarkets like US-American-Mexican HEB (Here Everything's Better), and Austin-headquartered organic store Whole Foods Market. Ethnic diversity also shapes the soundscapes of cities. Segregated housing areas are a good example. This sound segregation can occur as a result of voluntary exclusion, as in gated communities, but can also be caused by inner city (noise) barriers. Ethnic and social segregation in Austin, for example, manifests itself via city-dividing Interstate 35. The highway separates east Austin from the city's central area – including its downtown and university campus. East Austin traditionally is home to a large Hispanic population of lower social and economic status (a tradition only recently starting to be transformed by white middle-class gentrifiers). The extension of the downtown area with its vivid live music culture across Interstate 35 has not been successful, so far. The highway, with its loud traffic noise (see section 3.4.3 for sound level measurements), acts as an almost natural barrier.

The statement that "the more urban the visual setting, the 'less' pleasant and relaxing the perceived sound environment" may be too extreme (Viollon, Lavandier, and Drake 2002: 510). There are a lot of wanted urban sounds, such as those from live music events and venues, that add to the quality of life in a city. Overall, managing the different – sometimes competing – sound sources, and trying to suit every resident in a city, is an enormous challenge for today's city planning. As the following text will show, there are not only various sound sources but also several ways of using sound in the urban environment. So, the question is not whether they are perceived as being wanted or unwanted, but how they are used and dealt with.

4.2 Noise Perception vs. Noise in Urban Planning

The primacy of the visual in the urban experience is mainly distracted whenever sound occurs in an annoying or unwanted way. The sound of birds in a public park demands far less attention than construction noise, highway traffic during rush hour, and excessive live music performances at night. Thus, noise replaces sight and becomes the most important sensory input for the urban dweller (Adams et al. 2006: 2386, also see Bull 2000). That cities "are generally noisy places [...] is a taken-for-granted ingredient in our daily lives" (Atkinson 2011: 12). The observation of the urban environment as having a more negative soundscape than rural areas goes back to Schafer's acoustic ecology (1969). Explanations are rather vague. From a perceptional standpoint, technology sounds are considered less desirable than natural ones. Traffic noise in particular "is a growing issue" caused not only by cars and buses but also trains and aircraft (Greater London Authority 2004: 156). As discussed in 2.2.1, sound can be annoying and cause multiple health problems, from direct effects like hearing loss to indirect effects like sleep disturbance and cardiovascular diseases (Barrigón et al. 2010: 86). "Hundreds of social surveys of community response to transportation noise have been undertaken in the last half century" focusing on unwanted sound and trying to raise awareness regarding negative acoustic stimuli (Fidell et al. 2011: 791). Air traffic noise is most significant in areas close to international airports, while road traffic noise is generally described as "one of the prevalent environmental problems" (Law et al. 2011: 534).

Section 4.2.1 presents two different approaches to defining noise. Decibel measurements and individual perception are not always easy to combine, as 4.2.2 will show. The section describes how the city of Austin, Texas, marketed as the *Live Music Capital of the World*, deals with sound and especially noise complaints. Noise has been defined in every city code to manage the urban soundscape.

Meanwhile, the urban dweller's perspective is taken up in 4.2.3, in which we'll discuss how a new landing strip of Frankfurt Airport in Germany drastically effects the local population living in its landing path.

4.2.1 Noise – Definitions

Noise is a term heavily used in day-to-day life. Everyone seems to have an understanding of which sounds are to be labeled noise and which ones are not. Generally, the meaning is clear. Surely it depends on individual evaluation whether a certain music genre or a technology sound like a chopper's motor are considered noise, but generally the term seems to be detectable via the pure loudness of it: via decibels (dB). Environmental noise is defined as sound that is bothering and detrimental to one's health, caused by human activities, traffic, and industrial complexes (Bundesministerium der Justiz 1974: 32). This German approach from the 1970s can be found almost identically in the 2002 guideline on environmental noise by the European Parliament (see Europäisches Parlament and Rat der Europäischen Union 2002: 2). But:

> WHAT IS NOISE? Noise is unwanted sound. It is accordingly distinguished from signals, which are wanted sounds. In communication engineering when a message consisting of signals is transmitted, any sounds or interferences which impair its accurate transmission and reception are referred to as NOISE (Schafer 1970: 4).

The description of noise as 'unwanted sound' might not result in a precise and measurable value, but it does indicate that noise is something we do not like to hear. Deconstructing the soundscape into sound events, Schafer recognizes noise as "any undesired sound signal" (Schafer 1969: 17). A lot of authors follow acoustic ecology's understanding of the term when it is cited as being "undesired by the recipient" (Ingerslev 1972: 596), "an environmental stressor and nuisance" (Stansfeld and Matheson 2003: 243), or simply "unwanted sound" (Institute for Environment and Health 1997: 8, US Department of Transportation 1995: 3).

From a geographic perspective, noise can be described as "a sound which is out of place" (Atkinson 2007: 1905, taken from Gurney 1999: 6). This indicates that there is a certain way for sound to be in place, to fit into the urban environment, and to not disturb our hearing. To follow this logic, there have to be acoustic borders that define wanted sound as being within these borders, while noise is expelled from this certain order. What, from a phenomenological perspective, might sound like the alienation of noise – where the own also exists within certain boundaries and the alien is excluded – leads to legal justifications for what is perceived as being sound or noise. Enforcements and city codes generally refer to the subjectivity of such an evaluation: "the determination of 'unwanted' is very subjective and can vary substantially from one person to another" (Texas Department of Transportation 2011: 19). In the case of the state of Texas:

> the Federal Highway Administration (FHWA) has established absolute and relative criteria to more objectively determine when traffic sounds reach levels that result in impacts for humans (Texas Department of Transportation 2011: 19).

In addition to a relative criterion, where noise is detected as such whenever a certain decibel level is exceeded, there is an absolute criterion. "Under this criterion, a noise impact occurs when the predicted noise level approaches, equals or exceeds the [...] Criteria" (Texas Department of Transportation 2011: 19). Concrete decibel numbers always vary, not only by time of day and part of the city, but also by country and – possibly – state. The state of Texas – and this is important for the upcoming discussion on urban planning regarding sound in Austin – detects unreasonable noise "if the noise exceeds a decibel level of 85" (Texas Legislature 2007: 279) (section 4.2.2 will go more into detail regarding decibel measuring that always considers frequency).

Although cities try to define noise by decibel value and with reference to those sounds that are "an aversive component of a setting" (Jones, Chapman, and Auburn 1981: 54), one question remains: How does unwanted sound, sound? If it is defined via decibels, than why do some people struggle to stand in the front row of a live music concert? Why is it that some people are annoyed by the sound of children playing outside, while others love to hear those voices? To cover the individual element of noise perception, Austin, Texas has added a second definition of the term: "NOISE means sound which: (a) is louder than that permitted in this chapter, or (b) disturbs a reasonable person of normal sensibilities" (The City of Austin, Texas 1992).

Determining what "a reasonable person of normal sensibilities" is the big problem when it comes to figuring out a proper definition of noise. In regards to urban planning, it seems to be hard to follow this individual interpretation.

4.2.2 Noise in Urban Planning – Austin

Like every major city, Austin, Texas, has to deal with noise issues on a daily basis. Because of the high number of musicians and live music venues, the City Council created the Music Office to advise the city on related issues. David Murray, Sound Engineering Consultant for the City of Austin, defines the city's conception of noise "as something that is disturbing to people that are subjected to it" (Murray 2011). But despite knowing about the individual perception of sound and its labeling as noise, the city code cannot easily put this realization into practice. Noise regulations in the Austin City Code rely on decibels, not perception. Maximum decibel levels are defined according to time of day and neighborhood (see Cavanaugh, Tocci, and Wilkes 2010, Makarewicz and Zółtowski 2008, Texas Department of Transportation 2011). Generally, the Texas Legislature states: "[A] noise is presumed to be unreasonable if the noise exceeds a decibel level of 85" (Texas Legislature 2007: 279). For the downtown Austin area around 6th Street with the highest number of live music venues, 85 decibels – measured at the property line – is a value regularly exceeded during music performances. Joshua Leasure, working for a company dealing with sound insulation and a blogger for austinnoise.org, explains:

> Every venue that we stopped in front of we were at their property line was greater than 85 decibels. That's not surprising at all. It's a stupid number for that short of a distance. It doesn't make any sense at all. Everybody is always in violation of the ordinance, or almost always. So what's the point of it? (Leasure 2011)

Problems with the current city regulation started with the Texas Legislature's definition of the word 'decibel'. Frequency is missing in this description, which makes a profound difference in measurement results. Commonly, decibels are filtered using the weighting characteristic "A", where lower pitches are de-emphasized and primarily higher ones are recorded.

> It has been found that the A-scale on a sound-level meter best approximates the frequency response of the human ear (US Department of Transportation 1995: 3).

But some musical instruments also produce sound that can more effectively be measured using the C-scale, which, compared to the A-scale, is relatively unweighted. Using both values might be a better solution for determining levels of noise disturbance in urban areas with live music venues. "The frequency spectra highly influence sound quality" (Can et al. 2010: 768) and should not be reduced to a single scale. Cities like San Antonio, for example, differentiate between tonal and non-tonal sounds, those that are permanently heard and those occurring periodically. Adding or deducting up to seven decibels from a given sound limit because of acoustic differences makes a huge impact on the urban soundscape, as "an increase or decrease of 10 dB in the sound pressure level will be perceived by an observer to be a doubling or halving of the sound" (US Department of Transportation 1995: 4).

Dealing with sound and limiting it to certain levels goes back to the 1960s, when the "National Environmental Policy Act (NEPA) of 1969 provide[d] broad authority and responsibility for evaluating and mitigating adverse environmental effects" (Texas Department of Transportation 2011: 2). One year later the Federal Aid Highway Act involved "abatement of roadway traffic noise" (Texas Department of Transportation 2011: 2). Since then, regulations regarding noise haven't really changed. Noise ordinances are still written focusing on the source of the occurring sound. The perspective of the receiver is not considered in a city like Austin. "It doesn't think about the effect, only the cause" (Leasure 2011). Considering the harmful effects sound can have on the urban dweller (see 2.2) it is no wonder that noise complaints are getting stronger and that cities are having to figure out new ways for dealing with noise within the urban soundscape.

There are numerous reasons to complain about unwanted sound in a city. From dogs barking at night and heavy traffic to live music coming from music venues:

The urban environment is a noisy one. People seem to tolerate noise as part of the city, as they rarely try to improve the situation. "One argument is 'well you're living in the city, what do you expect'?" (Leasure 2011). There are situations when noise is simply not as pressing an issue as other sensory distractions, as the example of road traffic noise – "generally considered to be undesirable" (Blanco and Flindell 2011: 133) – illustrates:

> The people that live along there [close to Austin's MoPac expressway] got to vote on whether they wanted the [noise barrier] walls or not and they actually voted it down [...]. They felt the walls would obstruct their views (Leasure 2011).

Unlike in the European Union, where there are even regulations on how water-related projects should be built to protect people from negative acoustic input (Schäfer 2006: 5), in Texas there seems to be "a lack of [sensitization and] expertise" (Leasure 2011): "We don't have any comprehensive data for noise sounds, just complaints" (Murray 2011). Guidance to deal with noise abatement comes from states and the federal government (Kryter 2007: 1, US Department of Transportation 1995: 1). As discussed above, the restrictions are rather light and the guidelines often vague. Instead of reducing noise by transforming regulations and shifting from a source perspective to a receiver perspective, Austin tries to cope with noise complaints. In 2012 the 3–1–1 hotline was modified so that "residents are encouraged to dial 311 to voice a loud music complaint to a live operator, 24/7" (Austin Music Commission 2012: 3). The plan is to find out more about people's needs regarding the improvement of the urban soundscape before taking action. Interestingly enough the phone service focuses on complaints regarding live music, not noise in general. This seems strange – especially considering that the city's marketing is strongly related to music. Obviously, there are more sound sources in an urban environment than music. There is no department of the city's government that deals with noise, not even the Music Office, which concerns itself with organized forms of sound: "We're the closest but we don't [...] really deal with noise issues [...] so they pretty much fall off the map" (Murray 2011).

"There is no firm in Austin that deals with environmental noise" (Leasure 2011). As mentioned earlier (2.2.1) there are numerous ways to reduce noise. Automobile traffic is one of the biggest noise sources in a city. There are measurements, diagrams, and GIS-based models that try to find out about how noise is generated (Makarewicz and Galuszka 2011: 190). From tools to detect the number of heavy vehicles per road in relation to the time of day (European Commission Working Group Assessment of Exposure to Noise (WG-AEN) 2006: 63) to GIS-based models that "calculate [...] statistical noise levels" (De Coensel et al. 2005: 175), today there are ways to visualize urban noise not only three- but four-dimensionally (see Law et al. 2011: 534). Contrary to theoretical possibilities described in scientific papers, however, Austin's city code is based on older regulations that "from an acoustic standpoint [... don't] really make sense" (Leasure 2011).

> Urban acoustic researches are divided in three thematics: the sources characterization and identification, the acoustic propagation and the noise perception in an urban context (Richoux et al. 2010: 731).

All three topics seem to be at best only partially covered by Austin's current regulations. There is no doubt that "[m]easurement of noise in the field is difficult" (Kurakula 2007: 2) and "[u]rban traffic noise is known to vary highly" (Can et al. 2008: 1270), but as section 4.2.3 will show, there are possibilities to respond to urban noise. To begin with, every "city plan [...] should have a noise element", which Austin's does not (Leasure 2011). Generally, definitions of day-night average sound levels have to be included (Bundesministerium der Justiz 2006: 1, Europäisches Parlament and Rat der Europäischen Union 2002: 7, Fidell et al. 2011: 11). Urban dwellers have to be sensitized to noise and be aware of being a potential producer of it, as well.

One solution to the problem with noise in the urban environment is to handle existing emissions by putting up noise barriers (Banerjee and Southworth 1990: 564) "using concrete, wood, metal, [...] plastic", or earthen berms (Busch, Hodgson, and Wakefield 2003: 1947). In case of traffic, automobiles, trains, and aircrafts can also be re-routed to lower noise incidences (Ramis et al. 2003: 343). The best solution would be to avoid producing noise in the first place: "The redesign of the noise source, the vehicle itself, is a more fundamental way of attacking the problem" (Banerjee and Southworth 1990: 564). In contrast to most European cities, there are no building regulations in Austin that "specify sound insulation requirements for dwellings" (Rasmussen 2010: 373). Thus, the Downtown Austin Alliance, which tries to preserve and enhance the city's value – not only in terms of sound – worked on a solution regarding noise. As a different approach to the existing source-based model, the group focuses on the receiver:

> as they analyze the situation and say 'well we want to keep all possible receivers at 65 [decibels] or less'. [...] It's good for you neighbors because you know regardless of your location with respect to venues you're protected at that 65 level which is pretty loud for residential (Leasure 2011).

The lack of sound insulation on many buildings would be no longer a problem, as measurements would be taken not at the location of the source but at the location of receiver. Regulations on sound insulation could be a consequence of such a method. If a city like Austin wants to highlight its affinity for live music, venues should be able to prevent their performances from being diminished by sound regulations. To ensure a high-quality live music experience and also make sure people in nearby condominiums get their healthy sleep, sound insulation would have to be addressed. The argument that "[s]ound insulation design for buildings is an important task in environmental noise control" (Kurra and Dal 2011: 1) focuses on preventing sound from emerge from the inside. As long as the source-based model is in place, this does not really help people who receive and perceive sound as noise. But "that's really what it's about. What you're being exposed to, not the cause, but what the person is actually hearing. That's the number one thing" (Leasure 2011).

As the following section will show, noise perception always depends on numerous factors. To start learning about noise and to protect the urban population from too much sound impact, noise exposure maps are a good way of getting an overview of current sound levels within a city. In Austin, the only areas covered by such a map are Austin-Bergstrom International Airport and the Circuit of the Americas Formula 1 racetrack.

> In no large city has there been a systematic mapping of even the clearly definable qualities, such as [...] noise level (Banerjee and Southworth 1990: 91).

This perspective on US cities, given by Kevin Lynch in 1965, is still accurate for Austin in 2013. In the European Union noise exposure maps are required for every agglomeration and city of more than 250,000 inhabitants, every major road with over 6 million vehicles per year, every train track with over 60,000 trains per year, and every major airport (Bundesministerium der Justiz 1974: 33, Kurra and Dal 2011: 1). In Germany, for example, there is an even more specific addition to this regulation that describes how environmental noise has to be mapped (Bundesministerium der Justiz 2006: 1). Since 2006, traffic from road, rail, and air, harbor, industrial, and commercial zones has had to be displayed separately. All maps aim to transparently inform the public by providing materials in common data format and locations (Bundesministerium der Justiz 2006: 2–4, Europäisches Parlament and Rat der Europäischen Union 2002: 5). Regulations do not require empirical measurements, but rather a statistical estimate of how loud specific urban areas are.

> You have to do that predictively because if you're going to build a new road or railroad and it's not there yet you can't measure it. [...] You can go and try and find a similar environment and make measurements, but that's built in to the technique, the predictive technique (Leasure 2011).

Compared to the USA, in the EU there is far more sensitization to and research under way about how sound emerges in the urban environment and how it affects the local population (European Commission 2004). Thus, the following section moves away from Austin and focuses on noise perception using the example of air traffic noise in the Rhine-Main Area.

4.2.3 Noise Perception – Frankfurt

Both the perception and emission of noise are important for noise research. In the European Union the main focus concerning the latter lies with sound coming from transportation, i.e., road, rail, and air traffic, as well as "outdoor equipment"

(European Commission 2004: 4). The effects of noise are not only considered in regards to property values, especially in residential areas (Blanco and Flindell 2011: 133), but also with a focus on a population's health. Air traffic noise is an increasingly important issue in agglomerations with international airports. Entry lanes with relatively low heights of approach, in addition to sound caused by different air maneuvers, add to the existing unwanted soundscape. Cities like London have recognized the issue by pointing out the need to minimize:

> the impact of aviation noise [...] Issues involved in aircraft and airport-related noise management include: Quieter aircraft and aircraft numbers; Noise abatement operational procedures; Operating restrictions, including restrictions on night flights; Ground noise and surface access; Spatial planning and building insulation (Greater London Authority 2004: 146).

Acoustic ecology has long recognized the negative effect noise can have on the listener. Actions taken to eliminate unwanted sounds from the urban environment are important for enhancing quality of life in affected areas. From a phenomenological perspective, perception is not only related to the objects perceived but is based on an individual's ability to experience the world. Thus, general labeling of sound as noise has to be questioned. It has been mentioned before that sound perception relies on the listener's stream of consciousness. On the one hand, it is up to her/him if s/he thinks of sound as noise or not. On the other hand, high-volume sound can result in permanent hearing loss; noise can even be used to torture people (see 2.2). Painful volumes aside, physiologically "there is no distinction between sound and noise: sound is a sensory perception evoked by physiological processes in the auditory brain" (Berglund, Lindvall, and Schwela 1999: 23). Urban planning, though, is not paying too much attention to individual perception. Noise-exposure maps cannot display the single sound experiences of every urban dweller but can show decibel levels in areas like airports. Rising numbers of passenger and cargo flights make a strong monitoring of related sound emissions necessary. This is useful, as annoyance is found to be related to the sound level of aircraft noise (Miedema and Vos 2003: 1492).

> The effective (duration-corrected) loudness of noise exposure appears in most cases to provide a good account for social survey findings on the prevalence of aircraft-noise induced annoyance (Fidell et al. 2011: 803).

So, can individual perception be neglected when it comes to air traffic noise? Truax states that sound does have "a mediating effect" on the listener (2001: 94), but does that include the sound of airplanes, too? Some studies conclude that:

> [n]oise sensitivity has at most a very weak (positive) relationship with noise exposure, which cannot explain the strong influence of noise sensitivity on effects such as noise annoyance (Miedema and Vos 2003: 1503).

And there also seems to be a connection between the overall environmental sound level and parameters like the duration of time spent living with air traffic (Lim et al. 2008: 771). The following study was undertaken not to neglect the importance of a decibel-based model challenging air traffic noise but to highlight the individual perspective of that noise.

Frankfurt Airport is one of the largest European hubs for passenger and cargo flights. It ranks eleventh in the world in passenger numbers and ninth in cargo (including airmail) transportation. Located southwest of Frankfurt am Main, it is only about seven miles away from the city's downtown district and thus part of the Rhine-Main Area. The agglomeration, including important cities like Wiesbaden, Mainz, Darmstadt, and Frankfurt, is home to more than three million people. With its air traffic, Frankfurt Airport significantly adds to the region's soundscape. In late 2011 a new landing strip was put into operation to fulfill the airport's needs to grow and stay competitive. Due to an always-rising number of air traffic movements, people living under the aircraft's approach paths are confronted with a massive sound impact, generally perceived as noise. To find out about individual perception of sound emitted from planes, in 2012 a survey was carried out, including almost 3,000 questionnaires and 70 qualitative interviews. As Figure 4.1 shows, six research areas were selected, all directly affected by air traffic: Mainz-Bretzenheim (as part of Mainz), Hochheim, Bischofsheim, Rüsselsheim, Flörsheim, and Raunheim. With winds blowing east, aircrafts fly in from the west and pass over all six areas. Flight altitude significantly differs among them, from about 2,850 feet in Bretzenheim (which is still more than 10 miles away from the airport) to only about 1,090 feet in Raunheim (about three-and-a-half miles away).

Noise-protection areas are in place to ensure a maximum decibel load of up to 60 (see Figure 4.1), while sound emission of aircrafts selectively is much higher (see Flitner 2007). Official measurements are based on statistical data, using average values. With a maximum frequency of one aircraft per minute, emissions can significantly exceed the required maximum decibel level, because the area is much quieter the rest of the time. These regulations are part of the basic conditions that make it difficult for residents – at least temporarily – not to get massively affected by air traffic noise.

For the study, while questionnaires were filled out in a door-to-door survey to collect data from locals, quantitative interviews were aimed at key persons with special (noise-related or social) functions in each community. Interviewees were categorized into six fields: *politics*, *education*, *clubs and associations*, *citizens' groups*, *religion*, and *divers*. To avoid labelling the sound emitted from aircrafts as noise, both the questionnaires and the interviews started with questions on sound, moving to noise in general, and finally to air-traffic noise.

If existing statistical measured and source-based data were to be believed, nobody living in the Rhine-Main Area should have had a reason to complain about air traffic noise as long as protection areas were respected and sound emissions corresponded to emission values. The findings of the survey indicate differently.

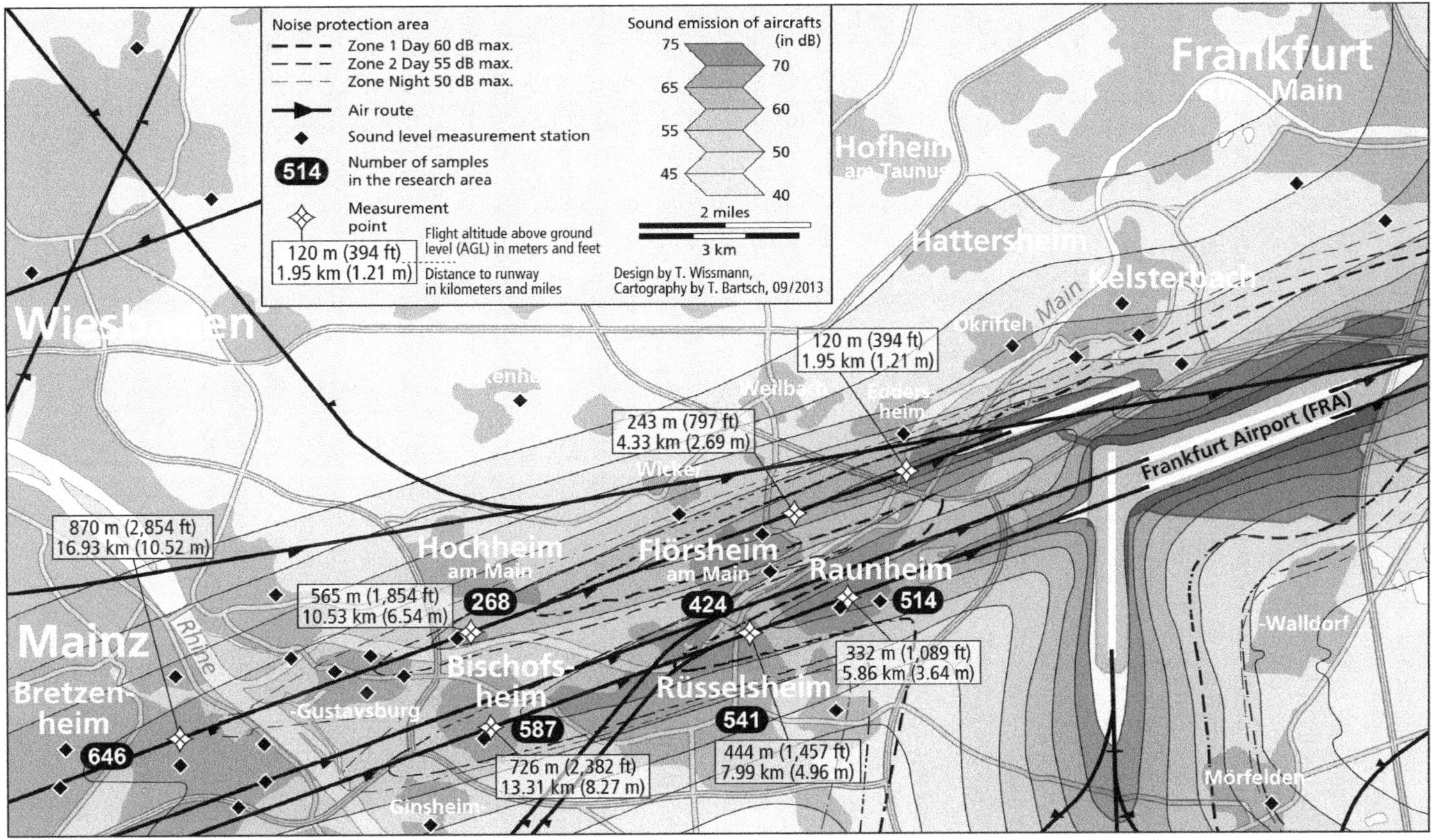

Figure 4.1 **Frankfurt am Main – Air-Traffic Noise**

People questioned display a representative sample of Rhine-Main Area's population. The sample size of 2,980 matches the general requirements (between 1,000 and 3,000 samples) in both Germany and the USA for meaningful data (Schnell, Hill, and Esser 2011: 269). The vast majority of respondents stated (84.08 percent) that there are sounds in their environment that disturb them. Technology sounds far outranked nature and human sounds in the category of "disturbing sounds", receiving 96.85 percent of the responses. The results match cities' countermeasures against urban noise, which generally focus on reducing or preventing traffic, commercial, and industrial sounds. Although there is a high density of autobahns and other major roads in the area, as well as a fully developed rail network, more than 90 percent of the disturbing technology sounds are attributed to aircrafts.

A first indicator for (air) traffic noise significantly co-shaping the soundscape and affecting it negatively occurs when respondents are asked about how they would rate the environmental sound volume. On a scale from 0 (total silence) to 10 (unbearable noisy) the average value is 6.46, with highest ratings for 7 followed by 8. More than 15 percent of the respondents rate environmental sound volume at 9 or 10, indicating massive impacts on their daily lives. As the survey was conducted after the opening of Frankfurt Airport's new landing strip Northwest, the perceived change of air traffic is interesting as aircraft movements are constantly rising. It is no wonder that more than 60 percent notice an increase, but that about 15 percent think aircraft movements have decreased is interesting. Crossing the results with the six different research areas help to explain the results. Respondents in Hochheim and Flörsheim, both north of the river Main (see Figure 4.1), as well as Mainz-Bretzenheim, describe an extreme increase of movements (up to 60 percent), while all three municipalities south of the river rank highest in *decrease* or *did not change*. Residents in Raunheim, closest to the existing landing strips, pick "decrease" as their top answer (over 36 percent). Overall, this reflects the change of approach paths in the area. Communities like Hochheim, Flörsheim, and Mainz-Bretzenheim receive more over flights today because of the new landing strip Northwest – Bretzenheim wasn't affected by air traffic at all until the end of 2011. As a result the loads of existing flight paths have been lightened, resulting in fewer overflights for the three other research areas. This difference in aircraft overflights and thus, air traffic noise emissions, is one reason for the different individual evaluations in the study concerning a future extension of the airport. Another reason is that many people living in the area are working for the airport or related businesses. Thus, answers are sometimes ambivalent and show a pull between annoyance and dependence.

> From an economic perspective, you have to say that Frankfurt Airport has to stay competitive. Otherwise a lot of people would lose their jobs. Thus, the new landing strip is great, as is Terminal 3, taking into account high total passenger transport. But I can understand everybody living nearby saying 'this is ridiculous'. It's really ambivalent (Balzer 2012).

In the questionnaires there is a general acceptance of Frankfurt Airport as an economic factor in the Rhine-Main Area. It is rated *most important* for jobs, international traffic, and the economy by the majority of all respondents. But about 72 percent of the respondents are against further development of the airport. There is strong dislike for the idea in Hochheim, Flörsheim, and Bretzenheim, while Bischofsheim, Rüsselsheim, and Raunheim are less opposed to airport development – in Raunheim, for example, more than 30 percent vote *in favor* (19.37 percent) or are *undecided* (11.07 percent).

> For Raunheim, the new landing strip is a positive thing. [...] There is an aircraft landing over Raunheim, while the following one flies over Flörsheim, but originally would have flown over Raunheim, too. Thus, the landing strip Northwest is a positive aspect (Zanetti 2012).

Aircrafts are recognized as a source of noise pollution by more than 90 percent of the respondents. While this seems to be a clear vote, not all people feel that air traffic noise impacts their daily lives. In fact, more people think it does not change anything (about 45 percent). Since landing paths depend on wind direction – Germany is generally mainly exposed to west winds coming from the Atlantic Ocean – landings over the research area are only taking place about one quarter of the time.

> We are extremely exposed to noise pollution between 5am and 11pm – if there is east wind. That's only true 25 percent of the time (Eckert 2012).

Given the fact that noise coming from the air only occurs temporarily and many people do not feel affected by it at all, its impact on the ones who feel disturbed is extreme. Those respondents said there were situations when people standing outside had to stop their conversations because of an approaching plane. Telephone calls were disrupted just the same. For them, it would become almost impossible to concentrate and do office work. Many windows remain closed even in the summertime – this despite the fact that almost no household in Germany has air conditioning.

> During overflights work is significantly harder, because you just can't concentrate. Calling somebody is impossible. Even with windows closed you sometimes have to interrupt a conversation. This is real stress (Winkelmann 2012).

As discussed in 2.2.1, sleep disturbance is considered the strongest non-auditory effect of noise (Kluizenaar et al. 2011: 1936, Ng 2000: 376, Smith 1991: 49). Thus, it is no wonder that most people said they are particularly annoyed by air traffic noise in the morning, between 5 and 8am (30.06 percent). Because of a ban on night flying, 5am is a considered a particularly bad time of disturbance as that is the time regular air traffic sets in. More than 35 percent of all respondents state that their sleep habits have changed because of the planes (a number that almost matches the about 40 percent stating that air traffic noise generally impacts their daily lives).

Most (about 60 percent) cannot sleep properly any more (Höfling 2012). Others take action and move their bedrooms to noise-protected places like the cellar.

> I'm woken by air traffic noise when the first plane breaks the silence at 5am. [...] Sometimes it feels like they directly fly into your house. [...] So, I wake up, cannot fall asleep again. I'm tired the whole day – and this is how many people feel (Jansen 2012).

Parents said they are especially stressed, as their children wake up way before their time (one mother of three started crying during the interview because she didn't know what to do anymore with her chronically overtired children).

The effects of noise pollution are seen across ages. Even in the most extreme situations, like prayer, silence cannot be guaranteed. Silence, which is supposed to help in the experience of a sense of sacredness and to support an inner communication with God (see 4.5), is disrupted by airplanes. "Last Sunday we had an open air service. Even with amplification you sometimes have to pause preaching because otherwise nobody would hear a thing. [...] That's it for silent prayer" (Winkelmann 2012). Other rituals are affected accordingly, as priest Sascha Heiligental illustrates:

> If you have to pause praying at a grave because a plane flies by and your words are masked, and people are disturbed in their most intimate feelings: That's outrageous! [...] There is a reason we talk about 'deathly silence'. [...] Air traffic noise clearly disturbs the services. A statement like 'Let us all pray in a moment of silence' is perverted by a plane thundering over you. [...] Think about baptism, when water runs down a child's head. Think about the blessing during a wedding. [...] It's unbelievable ... but then there is this helplessness ... (Heiligenthal 2012).

There doesn't seem to be a general decrease of quality of life as a result of air-traffic noise, but rather one that appears in different elements of daily routines. Over a third of the respondents feel like air traffic negatively impacts their leisure activities. The main issue concerns spending free time outside, like in one's garden or on one's balcony (65.78 percent). About 35 percent of all people state that they *agree* or *fully agree* with the statement that because of the aircrafts they cannot go outside any more. This matches almost exactly the number of people who said they stay indoors most of the time because of air traffic (about 32 percent).

> Quality of life is drastically reduced. I cannot use my garden any more, cannot have a chat outside any more. Just laying in the sun on your own property is practically impossible. Especially in summer you can sense the reduction of quality of life (Bröhl-Kerner 2012).

There are also people leaving the region to find a quieter place to live. "I myself got three friends who moved away from here because of air traffic noise pollution.

I think this is terrible" (Jansen 2012). Selling your house is not easy. Real-estate prices dropped significantly, and even those who had bought their homes 10 to 15 years earlier, still would likely owe money on its mortgage. Besides, those properties are not very attractive because of the noise pollution.

> I think I couldn't sell my house. Every possible buyer would be like 'What's going on here?' Of course air traffic noise reduces prices. Nevertheless, I know a lot of people who want to leave. [...] I have thought about moving away, too (Eckert 2012).

Some 28.51 percent of the population state that they have thought about moving, and 25.5 percent of homeowners thought about selling because of air-traffic noise. About 16 percent agree with the following statement: "Air traffic noise is so extreme that I just want to get out of here".

Still, not everybody wants to leave and some respondents said they actually appreciate the quality of life in the area. The study shows diverse perceptions about air traffic and related noise issues. Many people rate air traffic noise as far less significant than the quotes up to this point might suggest. The same argument we saw in Austin, Texas – "you're living in the city, what do you expect?" (Leasure 2011) – can also be heard in the Rhine-Main area: "You cannot have all locational advantages without noise" (Claas 2012). Regarding quality of life, some said the regions' biggest noise source is also one of its biggest assets: traffic. "Infrastructure brings us a special quality of life. Within two minutes you are at the train station, [...] we are only 15 minutes away from the airport, five to seven minutes to the autobahn" (Hartmann 2012). From an official perspective, air traffic noise is only one aspect among many. Patrick Burghardt, mayor of Rüsselsheim, rates quality of life in the city, especially related to noise, as follows:

> I am still convinced that Rüsselsheim has a high quality of life. This is no political talk. We have huge opportunities for local recreation. 65 percent of the municipal area is forrest. [...] Yes, there are parts in town where noise is overstraining. But I strongly believe that quality of life is very high, with everything we have to offer: Leisure activities, cultural activities, clubs [...] A few years ago we developed a new residential area that is completely sold out already. [...] We have two 'Central Parks' in town that are huge ... on a percentage basis you certainly could compare it to 'Manhattan' ... (Burghardt 2012).

How do people respond to the perceived noise pollution from air traffic? More than 27 percent state that they actively protest against it, mainly in demonstrations and through posters, stickers, and flyers pointing out the problems regarding aircrafts.

> Take a look around and you will find posters against air traffic noise at every corner. Some people are really engaged and inspire others who do not have the energy and time to help. This is really beautiful to observe (Reining 2012).

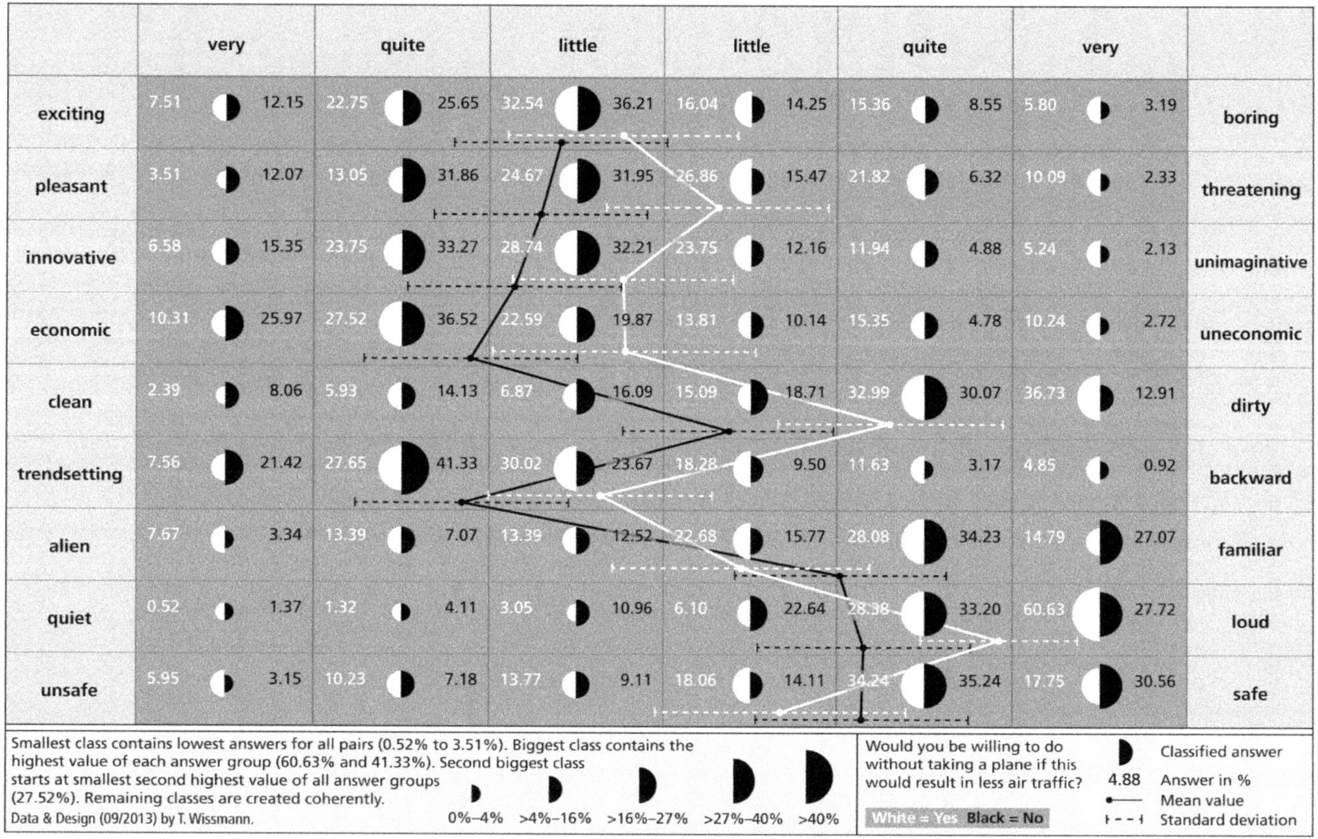

Figure 4.2 “What Do You Think About Flying?” (n = 2,550) (in %)

Protesting against air traffic noise helps to create a strong sense of community. Over 84 percent of the people in the Rhine-Main Area feel like they have a *good* or *very good* relationship with their neighbors. Among other actions there are very creative examples of how the population sticks together.

> People going to demonstrations together … I am pretty sure that this forges a bond between them. We have a singer-songwriter in town who composed a song against air traffic noise. The refrain was written together with people from the community (Müller 2012).

Although there are many forms of protest, there is also a sense "there might be not enough possibilities to discuss it" (Szafera 2012), and that nothing might change.

One way to lessen the impact of air traffic noise would be to reduce the number of aircraft overflights. Fewer planes means less related noise. Asked whether they would be willing to not take a plane to go on vacations in an effort to reduce the number of incoming and outgoing flights, the population is divided. While 41.4 percent stated that they would be willing to do without taking a plane, 47.54 percent said they wouldn't. In a Semantic Differential both groups were asked about what they associate with flying (see Figure 4.2).

There is a noticeable difference in all categories (on average about three thirds of a category). The most significant results concern the rating of flying as either *pleasant* or *threatening*. People who said they wouldn't go without taking an airplane draw a more positive picture of flying. The economic issue is the most favorable value for them, while the group of people who said they would go without taking a plane picked loudness as the most important factor. More than 60 percent of them described flying as *very loud*, by far the most extreme value of all answers – the second highest value received *only* about 41 percent. With its diverse answers the semantic differential shows that both perception of sound and its evaluation as noise are highly individual processes. The majority of a population may rate the sound of aircrafts as noise – and in some cases feel desperate about it – but even with aircraft movements predicted to rise in the future and noise pollution from the air getting stronger, there will always be different evaluations of air traffic sound. Whether people perceive it as annoying or not depends on their personal associations, experiences, and (physiological) perception. It seems that the best way to reduce your negative thoughts about air-traffic noise pollution is to take a plane for your next holiday.

4.3 Branding a City Through Sound

Arriving at Austin-Bergstrom International Airport is quite an acoustic experience. After leaving the gate, visitors pass by several shops, restaurants, and coffee bars to come to a stage, where live music is performed. This is the first hands-on contact with a city that brands itself as Live Music Capital of the World. What is advertised on shirts, mugs, and stickers can actually be listened to: live music.

Figure 4.3 Giant Guitars at Austin-Bergstrom International Airport

Luggage at the baggage claim circles around giant mockups of acoustic guitars (Figure 4.3). Before even leaving the airport, visitors are confronted with the meaning live music has for the city.

Aside from music there are various things visitors can explore in the capital of Texas – like the large urban bat colony under Congress Avenue Bridge – that Austin could use for city marketing. This wide variety and diversity is displayed by another Austin trademark, promoted by the Austin Independent Business Alliance: The slogan *Keep Austin Weird* (see Long 2010). But it is live music that the city has chosen to be associated with, including big music festivals, multiple live music venues, and a large number of famous artists.

The city's strategies regarding its brand will be discussed in section 4.3.1. The different live music districts will be uncovered and promotion and advertising will be discussed. Looking at how artists transport the live music image to the world we will show how important the local population is to keeping the brand alive. Section 4.3.2 will look at the relationship between the (live music) brand and the overall Austin population. There is no doubt that music is hearable almost everywhere in the city. Supermarkets sell selected CDs at checkout, and some children's playgrounds feature stages where bands perform on weekends. So the question is: Are bands being promoted actively to the locals. Is the general public asked to participate in the city's live music concept to keep the brand vital? One way to find out about the degree of activation is to take a look at popular media consumed by locals. Thus, the following discussion is based on a media analysis of the Austin American-Statesman, the city's daily newspaper. It will also be interesting to see if there is a particular music style or genre that is promoted by the city and associated with the slogan "Live Music Capital of the World".

4.3.1 Branding through Sound

In 1981 the City of Austin was given a new brand to market itself across the country and worldwide: *Live Music Capital of the World.* This moniker refers to the city's long connection to live music, associated with legends like *Stevie Ray Vaughan* and *Willie Nelson*, globally recognized festivals like Austin City Limits (ACL) and South By Southwest (SXSW), and the number of its live music venues. "Per capita we have more live music venues than any other city in the US or possibly the world", David Murray, sound engineering consultant for the City of Austin Cultural Arts Division, explains (2011). Since 1981 live music has been a successful way to raise the city's profile. With Rose Reyes, the local Austin Convention and Visitors Bureau (ACVB) even installed a "director of music marketing. It's a unique role in that we don't know of any other CVB in the country that has this full-time role within their CVB structure" (Reyes 2011). Both the City Council and ACVB work on promoting the brand. The Music Office, as part of the cultural arts division, was created in 2009, "in part to deal with sound problems, sound issues", arising from festivals, live music venues, and other sources (Murray 2011). Whenever there are noise issues, the Music Office mediates disputes between neighborhood groups and venues. The role of the AVCB concerning the brand, meanwhile, is more marketing related: "My role in the bureau is to incorporate music into as many aspects of what we do as possible […], to plug music in so that we can stand by our brand 'Live Music Capital of the World'" (Reyes 2011).

Other assets of Austin are also important in selling the city to tourists, like golf courses, local history, and North America's largest urban bat population – see the integration into audio-guided city tours in 4.4.3. But "music really kind of stands alone […] It's been a very successful brand in separating us from the pack, making us different and unique giving us something that people can come here and know its very much a part of who we are as a city" (Reyes 2011). Austin is the capitol of the state of Texas, but in contrast to many other major cities, it lacks several entertainment attractions. There is no amusement park like Disney Land and no major sports teams. But at night, music districts like Sixth Street and the Warehouse District provide numerous live music venues, and there are an estimated 200 annual festivals (Vallejo 2011).

> It is true that on any given night you can go out and hear a number of bands in each genre of music around town (Murray 2011).

> We've been able to create a story about our city that is really authentic. It's really about something that came together very organically and authentically and then somebody saw it and said this is why people come here. This is what people like. This is who we are, look around. This is who we are let's just call ourselves that. Embrace this part of our unique culture and use it to attract people to our city (Reyes 2011).

Position a city through its brand is more complicated than just sticking to it for more than 20 years. Austin has been able to get global recognition as an important location for live music because of the huge amount of marketing it has employed to transport the brand to potential visitors. In competition with other locations associated with music, like Detroit, Nashville, or Seattle, the city makes use of an amalgam of music-related features. In addition to its high number of music venues, festivals, and famous artists, Austin encourages musicians to identify with the city by offering them special benefits (see below). Thus, the thriving music community generates countable revenues.

> There's a total of 241.8 million [dollars] that are attributed solely to SXSW and ACL, the two biggest festivals. So it's going to be a considerable amount more because we have hundreds of music festivals (Reyes 2011).

Overall, the financial impact of the music business is "quoted as being one billion dollars annually. With a 'b', one 'billion'" (Murray 2011). In 2010, the music sector in Austin created about 1800 jobs and was the city's biggest creative force, outperforming film, gaming, and visual arts (TXP, Inc. 2012: 1).

Revenue is generated less from album sales than the general presence of music, thematically, all over the city. For tourists, the different music districts are points of attraction where they can experience live bands and directly spend their money. As Figure 4.4 shows, there are two major parts of the Austin downtown area where live music venues are located. About two thirds of the venues (46 out of 74) can be found in the area of Sixth Street, where there are up to 10 venues on one single street block – only about 350ft or 100m long. Especially during night hours the street is often closed for traffic, allowing pedestrians to experience music venues more freely. To further promote live music and generate additional areas for different music styles, the city is currently thinking about turning the area around Red River Street into a third music district (Reyes 2011). Locations in south Austin like South Congress Avenue offer additional opportunities to expand the live music scene.

The interesting question is whether this many venues can all compete. What about the large number of artists coming to Austin? Why is it so attractive for musicians to not only play gigs in the city once in a while but move there and label themselves as Austin artists or bands? Austin "just has so much to offer, so, to me, I still go to other cities and those cities don't have what Austin has as far as the health insurance and their own channels and stuff. It's still like very artist friendly, you know" (Vallejo 2011). Evaluating the situation for musicians in Austin is Alejandro Vallejo, director of the Creative Media Center, created by the Austin Music Foundation.

> Austin Music Foundation has been around for five years plus and they've always helped. They've done bootcamps for artists, and seminars and stuff. But usually they've done them like venues, like they rent out a club and have a bunch of people there and have a panel with producers and professionals. So, the Media Center was kinda put in place to have a place like even after you left the bootcamp that you could come back to and work on (Vallejo 2011).

Educating musicians to learn how the music business works is one of the assets Austin has to offer. The Media Center provides a rare service to unknown artists. Bands starting out often do not have the business-management background to live on the money they make with their music.

> The newer bands, they just think straight, and my job is to make them see like there's so many ways … there's so many things that you need to do, besides 'learn your ten songs and go play this club'. Because that's not it. That's, like, one part of it (Vallejo 2011).

Thus, locating oneself in Austin is less a question of local belonging than economic opportunity. If musicians know how to create their Web site, be present in social

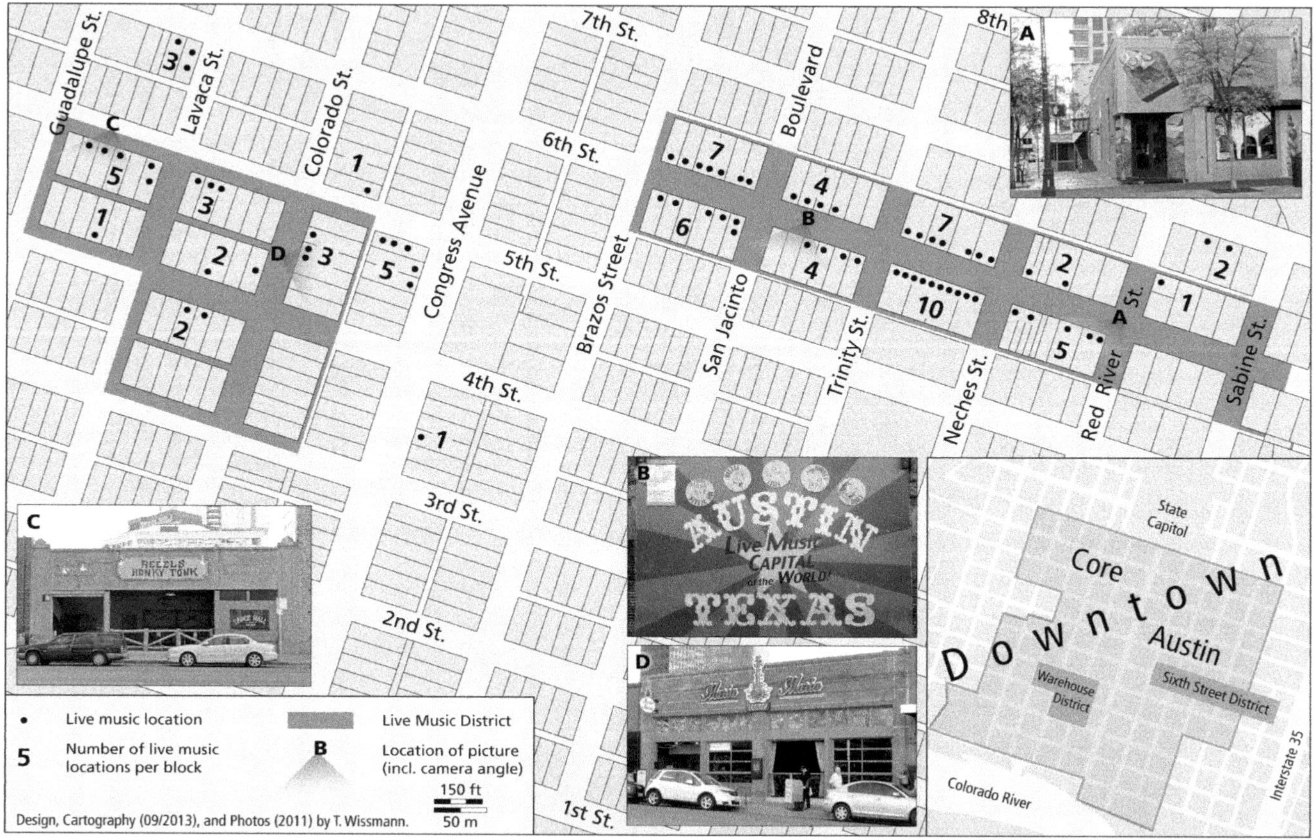

Figure 4.4 Live Music Locations in Downtown Austin, TX

media, sell records as well as shirts and stickers, if they earn some money from playing gigs, "if you do things right you can move ahead in Austin a lot quicker than some of the other bands" [in other cities]. [...] "A bunch of little checks make one big check. And in the music Business it is a lot of checks" (Vallejo 2011).

Local artists can also become members of the Health Alliance for Austin Musicians. Its mission:

> is to provide access to affordable health care for Austin's low income, uninsured working musicians, with a focus on prevention and wellness (Health Alliance for Austin Musicians).

The possibility of getting health care is a strong consideration for many artists thinking about relocating to the city and henceforth calling themselves Austin artists.

Bands also gain because of the city's thriving local music scene. As stated above, the city is promoted via its festivals, venues, and also its musicians. Whenever a band becomes famous, a trail in its biography leads back to the *Live Music Capital of the World* slogan. "We also work directly with some Austin artists that are on tour. When they're on tour often times they will invite people to come to Austin on our behalf" (Reyes 2011). These musical ambassadors represent the city at big festivals like Lollapalooza or the American Music Festival. They are also sent to Austin's sister cities all over the world to "help promote Austin's image abroad" (Murray 2011). Musicians in town are also included to help market the brand. "We have partnerships with the House of Songs. It's an Austin-based organization that pairs international songwriters with Austin songwriters here in Austin in a physical house. They write songs" together (Murray 2011).

In the city itself music is integrated into as many events as possible.

> We have a big swimming tournament that is taking place at UT [University of Texas] next week. For the second year in a row we've hired a live band that actually performs from the diving boards while some of the swim races are taking place (Reyes 2011).

The above-mentioned live music stage at the airport also promotes the brand within the city, as do music events in City Hall. Hotels have flyers in their rooms featuring information about the music scene and the different music districts. Media is used to promote the brand, too. The city's own Web site includes a section where visitors can listen to typical Austin music. An annual compilation of songs can be ordered online. Promotion on music-related Web sites and journals is joined by the use of media that only indirectly cover music, like travel magazines and food publications.

> We host journalists worldwide, year round. A lot of times it might be a food publication that is coming to town but in the course of showing them our culinary

> scene we also take them out to a live music venue. That inevitably works its way into the story (Reyes 2011).

Exporting the brand can also work with entertainment products, such as the Dell mp3 player – the company is headquartered in nearby Round Rock – that was labeled "made in the Live Music Capital of the World" (Reyes 2011). Connected to SXSW, the Music Office has an event called Live Music to the World, where:

> we're going to get a shipping container and fill it full of musical instruments and recording equipment and send it to either Nairobi, Kenya or Haiti. [...] It's a philanthropic gesture to help other countries in need but also to help strengthen the Austin brand (Murray 2011).

The promotion of the brand has been very successful and many people associate Austin with music. But what kind of music? Is there a typical genre for the capital of Texas like the grunge of Seattle, country of Nashville, and techno of Detroit? In the early days of the brand and even before that in the 1970s, "it was a cosmic country thing. *Willie Nelson, Waylon Jennings* and all those guys, and then it was the blues and *Stevie Ray Vaughan and the Fabulous Thunderbirds*, and they had a tremendous impact on Austin" (Murray 2011). In the following years the genre did not seem to change, as the "significant style in the late 1980s and 1990s" has been described as blues rock, or Texas blues (O′Meara and Tretter 2013: 65). Today, the acoustic guitar of rock, blues, and country artists still seems to be the most important instrument:

> Austin, I think, is in the Americana music movement. Austin would be considered roots music. There's hard rock too. [...] the Austin sound would be an amalgam of all those. [...] [A]coustic, rock, blues that's what I think of when I think of Austin (Murray 2011).

Due to a constantly changing music scene, new genres became more important, as Rose Reyes (2011) notices:

> Our Indie rock scene, our Punk scene, has really grown. I would say that's definitely one of the main genres. I think for a long time people associated Austin with roots music. Blues like Stevie ray Vaughan kind of sound or even the country sound like Willie Nelson. Those roots music forms are still very prominent and a big part of our identity as a music town but like I said the Indie rock scene has worn a lot and also the Latino music scene has grown a lot (Reyes 2011).

This quote is really remarkable, as the Austin music scene is displayed as being rich and diverse. In connection with the city's brand this variety does not seem to persist. Every year ACVB releases a CD compilation featuring Austin artists to let listeners become familiar with the typical Austin sound. An analysis of volumes 6 to 10 of the series shows that more than one third of the songs listed are rock songs. The second third goes to country and folk in equal parts, leaving the last third to pop, R&B/soul, jazz, and funk (see O′Meara and Tretter 2013: 67). What might be interpreted as "ignore[ing] the city's racial, ethnic, cultural, and even musical diversity" (O′Meara and Tretter 2013: 53) could also be explained with a certain need to consistently develop the existing brand, which has been defined by the acoustic guitar from the start. Another explanation for ostensibly (almost) ignoring genres like Latin in promotional material like the ACVB's CD series could be that many Austin artists – irrespective of their cultural background – define their music not as Latin but rock, punk, or independent. "[E]verybody thinks Latin music is like Salsa and Merengue and it's not. For every band America has we have the same. We have a U2, there's a punk rock band, there's all these different styles of music instead" (Vallejo 2011).

The fact is that festivals like SXSW and ACL are heavily promoted and well-known internationally, while new events like the Latin music-focused Pachanga Fest – which only started in 2007 – are also being supported and marketed in context with the city's brand. Whether or not there are other genres that might be mentionable to display the live music culture of the city more accurately, the main focus for the brand is Austin's traditional rock, country, and blues. The latest CD release (volume 11, in 2013) of the ACVB series finds the established music mix jazzed up with some elements of punk, independent, and Latin. The following chapter (4.3.2) will come back to the issue by looking at the appearance of different music genres and local artists in the local newspaper. The definition of 'live music' is also significant for the understanding of the brand.

> [W]hen people think of live music they think of people playing instruments. I don't want to list DJ bars and they show up hoping to see some Texas music and it's just a DJ playing music that they could hear anywhere in the world (Reyes 2011).

As the city tries to attract visitors to come experience Austin because of its live music scene, there is a focus on audiophiles. "[W]e do try to target audiences that we know that already love music and maybe haven't been to Austin" (Reyes 2011). How many tourists visit Austin for its music is hard to determine. Even if a visitor's primary reason is not the music, inner-city promotion like the stage at the airport and flyers at hotels (see above) attempt to attract every visitor. And if music is the reason why people come to Austin, it supports other assets of the city. It "draws people here and then when they're here they discover our visual arts scene. They discover other parts of our cultural arts scene" (Reyes 2011).

With tourists alone, however, the numerous live music venues could not survive. The city also needs locals to attend festivals and go to the bars and clubs at night. Close to 50 percent of the attendees of ACL are locals, for example. "Local residents have also maintained a consistent support for music and other cultural events" (TXP, Inc. 2012: 8). Asked about the identification of Austinites with the brand, officials are confident that everybody knows about it. "I think it's kind of a given that locals know about the scene" (Murray 2011). That does not mean that everyone living in Austin identifies with music or even the brand, but that it is a positive connotation to be proud of: "My guess is that they're going to have some pride about that even if they're not personally music fans or they're not personally involved in the live music scene" (Reyes 2011). But there is no organization or official office that pitches the brand to the urban dweller. "As far as the brand itself I would say no" (Reyes 2011). ACVB does not do it "because our mission is solely tourists and visitors" (Reyes 2011). The Austin Music Foundation is caring about local artists "to make sure that bands are making money and people are getting information and they can use to go producer record, play at a club and get paid" (Vallejo 2011). And the City of Austin Cultural Arts Division deals more with noise complaints than music marketing, although promotion to locals without doubt seems to take place: "I don't know who [does it], but it's on coffee mugs and bumper stickers. [...] [I think it's] vendors" (Murray 2011).

Live music, as the city's biggest marketing asset is believed to be "our bread and butter in Austin. If we as citizens don't appreciate it we could lose that as part of why people are interested in Austin". (Reyes 2011) Without music, the city's sense of place would radically change. As an economic but also socio-cultural factor this organized form of sound is important for the urban dweller. As the following section will show, media, both by generating and expressing public opinion, tells about the impact live music and sound – especially when used as the city's brand – can have on the local population.

4.3.2 Impact of Live Music and the Sound Brand

In a city like Austin music is all over the news. Local media reports daily about upcoming events, festivals, and live music performances. The urban dweller is confronted with the city's live music asset whether they like it or not. As discussed before, local artists can get in touch with organizations like the Austin Music Foundation to learn more about how to survive in the music business. Austin Music People, a group formed in 2010, brings together neighborhood associations, music venues, and developers, musicians, music industry, and music fans:

> making the economic development of the Austin live music industry a priority for our political, civic, and cultural leaders, and to connecting and empowering Austin's live music community by providing powerful advocacy tools,

> participating in policy development, and representing live music interests in public and private forums (Austin Music People).

In contrast to supporters of the Live Music Capital of the World brand, like local artists, the general public is confronted more casually. This section will show how the general public learns about music over the media. Articles like those in the daily newspaper Austin American-Statesman, encourage people to attend local festivals or support live music by going to a club at night. What kind of music they are told about strongly concurs with the Austin's image as a Texan rock, country, and blues town. A comparison of mentioned genres and talked-about local artists also reveals a tendency to support the city's brand in local media. How much the local population is confronted and how far they are integrated into the live music economy can be illustrated by taking a look at organizations that both organize and support Austin's music events. Even though the urban dweller is not addressed directly by the marketing machinery of the Live Music Capital of the World brand, in a city like Austin there is little chance to escape this predominant sound theme.

The Austin American-Statesman (AAS) is Austin's major daily newspaper. In addition to the alternative Austin Chronicle, which is more tabloid-style and distributed for free, the Statesman is the most important print medium in the city. With a circulation of about 120,000 daily (about 180,000 on Sundays), it is currently ranked 62 in the list of the top US daily newspapers (Alliance for Audited Media). To find out how the live music brand and the live music scene as a whole are covered in the AAS, a quarter-year sample was analyzed. From September 2011 to November 2011 all 91 issues (daily and Sunday) were collected. First the complete texts – excluding ads, headlines, and captions under illustrations – were worked through using 15 "hotwords", i.e., music-related terms. Table 4.1 shows that 236 articles contained one or more hotwords. Overall nearly 1000 entries were found. The most frequently used hotword was *band* followed by *festival* – both in total number of articles and single entries. Whenever a term occurred it was counted, even if it was used multiple times within on article. In addition to musical terms in a narrower sense – like *singer*, *concert*, and *band* – general terms related to the current study were noted, like *soundscape, brand,* and *Live Music Capital of the World (LMCotW)*. Except for *soundscape*, every term occurred at least once in the analysis, although *live performance, sonic, brand, acoustic,* and *LMCotW* were mentioned only sporadically.

Hotwords were only counted if they appeared not as part of another word, like *band* in hus*band*. They were also excluded whenever they had another meaning, like the use of *sound* as a verb instead of a noun. Plural terms were included whenever necessary, while all appearances of words as part of a band names like *High School Singer* were not considered. The term *music* was not used as a hotword because of its general use and common occurrence, which would have

Table 4.1 Austin American-Statesman – Hotwords Sound and Music

	September		October		November		Total	
Hotwords	**Articles**	**Entries**	**Articles**	**Entries**	**Articles**	**Entries**	**Articles**	**Entries**
Acoustic	0	0	2	2	3	4	5	6
Band	29	145	23	50	29	92	81	287
Brand	2	3	1	1	0	0	3	4
Concert	19	36	15	40	12	27	46	103
Festival	49	135	6	16	12	39	67	190
Live music	17	23	19	27	19	22	55	72
Live performance	0	0	1	1	1	1	2	2
LMCotW*	2	2	1	1	4	4	7	7
On stage	7	9	4	4	2	3	13	16
Performing	9	13	9	11	2	2	20	26
Singer	18	24	21	30	26	32	65	86
Songwriter	5	8	8	9	13	15	26	32
Sonic	2	3	0	0	0	0	2	3
Sound	21	43	13	22	20	28	54	93
Soundscape	0	0	0	0	0	0	0	0
Total	**180**	**444**	**123**	**124**	**143**	**269**	**446**	**927**

* Live Music Capital of the World

led to a less meaningful sample of articles, but was reconsidered in step two of the analysis.

To focus on the articles only relevant to the Austin live music scene and brand, all articles were filtered out that were not related to the capital of Texas. Thus, events like the Harvest Music Festival in Arkansas in October 2011 were excluded from the analysis.

Some 202 articles were further investigated in step two of the analysis, this time counting hotwords within Austin-related texts,. After importing all documents into MaxQDA software for qualitative research a narrow context analysis was performed. To find out how the local population is involved with live music and the brand, five categories containing 12 codes were created: *location*, *event*, *activation*, *genre*, and *player*. The last contained eight coded subcategories, while the first four were only categories and codes. A total of 1,946 codes were assigned.

Location signifies all places in Austin where an event takes place. The location has to be related to music, like the Frank Erwin Center, a multipurpose arena where live music concerts take place. The code appeared 369 times – more than any other one.

Event can be a concert in a live music venue, a festival like ACL or SXSW, or similar. It has to take place in Austin. It was assigned 197 times.

Activation signifies all parts of a text where the reader is addressed to participate in a music event. This can be a plain announcement of an upcoming live concert, where only one line mentions the date, time, and location of the event. A stronger activation would be an explicit invitation to join a relevant concert or festival. The code occurs 167 times, and does not only address potential visitors, but also donors and other supporters of the live music scene.

Genre highlights all different music genres that appear in relevant articles. With 163 mentions, the term appears almost as often as *activation*, only the explicit naming of a genre like *rock*, *country*, and *blues* is counted. Reports about a certain artist or band that implies a genre – like country for the artist *Willie Nelson* – is counted in one of two different *artist* subcategories, such as *player.*

Player includes all relevant people and organizations that appear in context with live music and the brand. It is by far the biggest category, as it consists of eight single codes. While locations and events from out of town were neglected, non-local player are considered of importance for Austin's live music scene. An artist performing at *ACL* might be from another city, state, or even another country, but there undoubtedly is a connection to the city via the festival where s/he performs. Thus, artists are divided in *local* (240 mentions) and *non-local* (365), as are *organizers* and *suppliers*. The former describes an organization or private investor that supports live music and/or the city's brand by arranging a related event or helping the music scene to thrive. An example is Austin Music Foundation, which cares about the career of local artists by organizing live gigs, workshops, and bootcamps. Local organizers are far more reported-on (154) than non-locals (35). Whether more local organizers invest in local music or whether the newspaper simply reports more frequently on those from Austin cannot be said. But it seems to make sense that local patriotism plays some kind of role. Like organizers, local suppliers supporting a music event by catering for or reporting on it are way more integrated into Austin's music scene (105) than non-locals (24). The last two codes were applied to *visitors*. In contrast to all other players, non-local visitors could not be detected, and there was only a relatively low number of local ones (16). A major part of the visitors appearing in the coded articles could not be classified by origin. Thus, most of them (110) are listed without information.

To learn more about the single codes, step three of the analysis adds weighting and more detailed information. Whenever a code occurs, narrow context analysis reveals its context and importance for live music in Austin. *Location* is not weighted but labeled as local or non-local in step two. A mandatory quality of location is its attachment to live music, making any further differentiation unnecessary. The *event* taking place at a location is weighted in three categories. If some live music is played at a local farmers market, for example, it is not the predominant theme of the event and is, therefore, given a single rating. Events like those center around paintings, sculptures, installations, and other arts are double rated. They cannot exclusively be labeled music events but have music as their main focus. A triple rating is applied to events that are solely about (live) music. The newspaper article does not have to be long, but the focus of the event itself is essential to its weighting, as the following example for a triple weighting shows:

> Fun Fun Fun Fest experienced a 40 percent growth in attendance over last year's festival, with an average of 15,000 people a day over the course of the three-day event earlier this month [...] (Mongillo 2011a).

As described above, data about the origin of visitors is hard to extract. Nevertheless they are ascribed three different weights in the analysis. If visitors are mentioned within a relevant article but do not attend a music event, the rating is only single. A double rating represents their being mentioned during a music event without a closer description of what they are doing. "Dust brought on by dry and windy conditions didn't seem to bother most people filing into Auditorium Shores on Saturday for day two of Fun Fun Fun Fest" (Mongillo 2011b). A triple rating, as the highest weighting for visitors, focuses on them and their actions during a music event, or even contains individual portraits: "My wife and I have enjoyed seven of the 10 Austin City Limits festivals. This year, as always, we bought three-day passes as soon as they went on sale" (McDonald 2011). Comparing explicitly local visitors with undefined ones, weighting is almost identical. The data suggests that there is no special promotion of locals attending live music events. Even though about half of the codes were weighted the highest, it seems that Austinites are not explicitly addressed through the newspaper.

More information about marketing live music to the local population through media can be acquired by analyzing the *activation* code. In analogy to the former codes a triple weighting tells about the intensity with which people are invited to participate in a live music event. The results show almost no effort on the part of the media and players like ACVB to promote live music to the local population. More than half of all passages coded with activation are of the lowest intensity. To make sure all efforts of the music industry communicated via the Statesmen were captured, even pure announcements like the following were integrated into the analysis, counted with single weighting: "*The Intergalactic Nemesis* – When: 8 p.m. Tuesday – Where: Dell Hall, Long Center, 701 W. Riverside Drive – Cost: $19–$49" (Van Ryzin 2011). The reader of the newspaper is addressed by this announcement, but it is far from promoting live music or even the city's brand. Articles receiving double weighting are not significantly more active in pitching live music:

> The month in music kicks off in grand fashion Wednesday at the Long Center, when a host of musicians, including Lyle Lovett, Verlon Thompson, [...] and many more celebrate the 70th birthday of songwriting legend Guy Clark (Mongillo 2011d).

Only items extensively promoting artists or music to the reader, or explicitly inviting them to join an upcoming event, receive triple weighting:

> Conspirare Big Sing. No experience necessary for this concert where the audience is the choir. Conducted by Craig Hella Johnson and hosted by the Conspirare Symphonic Choir. Vote for songs to sing on Conspirare's Facebook page before the event. 6 p.m. Thursday. First Baptist Church, 901 Trinity St (Martinez 2011).

There is no stronger form of activation than the approximately 37 percent triple weighted passages. Originally, there was quintuple grading set up for all codes, but – except for *local artists* – the tripartition was sufficient enough. There is no stronger activation in the Statesman to get locals to join live music events that would have required more extended weighting. Informing the public about upcoming events can hardly be called promotion of the live music scene. But newspapers can inform readers about local organizations and private investors that help the brand and music scene to thrive by supporting events and organizing them.

In all music-related articles there are 84 organizations mentioned involved with events in Austin. More than half of them (45) can be described as *organizers*, as they host, arrange, or present a live music event in the city. The other half

Table 4.2 A.A.-Statesman – Organizations Involved With Music Events

	Organization				Organizer		Organizer (mentions)			Supplier		Supplier (mentions)		
Area	**Total**	**In %**	**Weight**	**Weight %**	**Total**	**In %**	**Total**	**Weight**	**Weight %**	**Total**	**In %**	**Total**	**Weight**	**Weight %**
Art	11	13.10	29	5.73	2	4.44	3	8	2.66	9	23.08	12	21	10.24
Education	8	9.52	82	16.21	5	11.11	24	57	18.94	3	7.69	11	25	12.20
Event management	16	19.05	181	35.77	10	22.22	43	89	29.57	6	15.38	41	92	44.88
Financial	1	1.19	1	0.20	0	0.00	0	–	0.00	1	2.56	1	1	0.49
Food	9	10.71	13	2.57	7	15.56	7	8	2.66	2	5.13	2	5	2.44
Government	4	4.76	26	5.14	2	4.44	7	12	3.99	2	5.13	11	14	6.83
Health	9	10.71	42	8.30	4	8.89	13	31	10.30	5	12.82	7	11	5.37
Hotel	2	2.38	2	0.40	2	4.44	2	2	0.66	0	0.00	0	–	0.00
Media	7	8.33	76	15.02	4	8.89	27	54	17.94	3	7.69	14	22	10.73
Real estate	3	3.57	6	1.19	2	4.44	3	5	1.66	1	2.56	1	1	0.49
Retail	1	1.19	1	0.20	0	0.00	0	–	0.00	1	2.56	1	1	0.49
Social	12	14.29	46	9.09	6	13.33	24	34	11.30	6	15.38	7	12	5.58
Transport	1	1.19	1	0.20	1	2.22	1	1	0.33	0	0.00	0	–	0.00
Total	**84**	100	**506**	100	**45**	100	**154**	**301**	100	**39**	100	**108**	**205**	100

(39), named *supporters*, support the events with different supplies like food, additional artistic programs, or music equipment. In a two-step generalization they were divided into 11 categories each (see Table 4.2). Organizations like the Health Alliance for Austin Musicians were merged into the group *insurance companies*, for example, added to the *health* category. Local radio stations like KGSR were summed up in the *radio* category, which was later merged with TV stations, print, and online media to form the *media* category. As Table 4.2 shows, there is a difference between the total number of organizations participating in live music (*total* columns) and the intensity with which they are involved in an event (*weighting* columns). In terms of variety, *event management* is followed by *social* and *art* organizations. There is a high number of *food*-related organizations putting on music events, and *art* organizations are very present as suppliers. But despite the pure numbers, the most important and most involved organizations are associated with *event management* (35.98 percent), *education* (16.3 percent), and *media* (15.11 percent).

Like with the other codes described above, a three-way weighting is applied to all organizations. Single weighting is used to display mentions that are of only minimal relevance to the music event, while double weighting implies a stronger attachment to an event or artist. Triple weighting is awarded to organizations that are portrayed as strongly supporting Austin's live music scene:

> Here and there, the visitor could spy the captains of Austin's music industry, delegates from key groups such as C3 Presents, South by Southwest, ACL Live, Austin Music People and Transmission Entertainment (Barnes 2011).

Although there are a high number of social and art organizations involved with live music, they do not have an equivalent impact in the newspaper. Instead, *Media*, while only ranking seventh in mentions, is the third most important category according to media impact. The high impact of *education* can be explained by the heavy presence of the University of Texas (UT) both as organizer and supplier of events.

> The concert, organized by area musicians, the University of Texas, […] [and others], will raise money for victims of the wildfires that destroyed more than 34,000 acres and 1,600 homes in Bastrop County this month (Mongillo 2011c).

Radio and television stations are located on campus, artists are attached to UT, and music venues are also to be found there. Thus, it is really the music scene itself that profits from live music in the city. Especially *event management* – with a media impact of over one-third alone – is the outstanding profiteer off the press coverage. The music industry primarily seems to support itself, while local non-music groups profit less. Suppliers like food trailers from the local food industry might be present at many music events, but at least in the Statesman they do not play a significant role as part of Austin's live music ecosystem.

Table 4.3 A.A.-Statesman – Occurrence of Music Genres (September–November 2011)

	Music genre associated with local artist				Explicit mention of music genre			
Genre	Total	In %	Mentions	Mentions %	Total	In %	Mentions	Mentions %
A capella	4	–	1	0.23	–	–	–	–
Acoustic	24	1.93	9	2.04	–	–	–	–
African	3	0.24	1	0.23	3	0.81	1	0.43
Alternative	56	4.51	27	6.12	–	–	–	–
Ambient	3	0.24	3	0.68	–	–	–	–
Americana	32	2.58	12	2.72	8	2.16	4	1.71
Blues	120	9.66	33	7.48	10	2.70	7	2.99
Children	5	0.40	3	0.68	–	–	–	–
Classic	94	7.57	25	5.67	20	5.39	15	6.41
Country	107	8.62	36	8.16	35	9.43	19	8.12
Electro	29	2.33	9	2.04	20	5.39	14	5.98
Folk	59	4.75	26	5.90	18	4.85	11	4.70
Funk	33	2.66	10	2.27	14	3.77	8	3.42
Hip-Hop	15	1.21	5	1.13	18	4.85	9	3.85
House	3	0.24	3	0.68	–	–	–	–
Indie	66	5.31	31	7.03	16	4.31	13	5.56
Jazz	53	4.27	12	2.72	15	4.04	11	4.70
Latin	19	1.53	6	1.36	5	1.35	2	0.85
Marching	5	0.40	3	0.68	–	–	–	–
Metal	8	0.64	1	0.23	10	2.70	8	3.42
Oldies	–	–	–	–	5	1.35	4	1.71
Pop	88	7.09	38	8.62	22	5.93	13	5.56
Psych	11	0.89	7	1.59	2	0.54	2	0.85
Punk	35	2.82	19	4.31	34	9.16	19	8.12
R'n'B	28	2.25	7	1.59	5	1.35	4	1.71
Rap	14	1.13	2	0.45	12	3.23	8	3.42
Reggae	2	0.16	1	0.23	4	1.08	2	0.85
Religious	4	0.32	1	0.23	13	3.50	9	3.85
Rock	237	19.08	82	18.59	61	16.44	35	14.96
Rock'n'Roll	15	1.21	10	2.27	–	–	–	–
Soul	70	5.64	18	4.08	19	5.12	15	6.41
Swing	–	–	–	–	2	0.54	1	0.43
Total	1.242	–	441	–	371	–	234	–

As discussed above, the local music scene is diverse, but only some genres are used to brand Austin as the *Live Music Capital of the World*. If officials do not actively promote the brand to the local population, the local music scene might have a different focus on what kind of genres are played in live music venues.

The media analysis of the Statesman shows the drastic impact the brand and its associated music have on the Austin music scene. The ACVB promotes a variety of genres: "We have jazz, we have blues, we have rock, we have Latino, soul, we have hip-hop and so there's something for all of you attendees" (Reyes 2011). But the brand's preference for rock, country, and blues is clearly transported through local media. The categories *genres* as well as *local artists* listed in AAS are both used to find out about the importance of music genres. As Table 4.3 shows, the most important genres are rock (16.44 percent), followed by country (9.43 percent), and punk (9.16 percent), while blues receives only 2.7 percent of the attention. Though this implies an – at least partial – advance of newer genres, the total mentions of genres and a comparison of local artists tell another story. *Rock* is by far the most covered genre in the Statesman. It accounts for about 15 percent of the total mentions, and is weighted even higher (16.44 percent). This means that articles on *rock* not only mention this genre but talk about it in some detail. A look at *local artists* shows that those who play rock get even more attention, as they are mentioned even more often (18.59 percent). There are portraits and in-depth descriptions, leading to an overall 19.08 percent weighting of *local rock artists*. A similar increase is detected for *country* and related *local artists*. The analysis of the importance of punk in the Statesmen tells another story. While the genre accounts for 8.12 percent of all *genres* mentioned, and receives even stronger attention regarding weighting (9.16 percent), only 4.31 percent of the *local artists* mentioned in the newspaper play punk. The result for weighting is even more dramatic, as the importance of the media coverage of punk artists is as little as 2.82 percent. Compared with the traditional Austin genre blues, it becomes clear that the *genres of the brand* are favored by the local newspaper that reflects – intentionally or not – what can be heard on every promotional CD: not only rock and country, but also blues still dominates the music scene when it comes to media coverage. The 2.7 percent weighting of the blues genre shows, connected with a slightly higher percentage of mentions (2.99 percent), that – according to the Statesman – there is very little blues happening in Austin. But local blues artists are of much higher importance, as they are mentioned 7.48 percent and get even more intense coverage (9.66 percent).

Overall, support for live music is obvious. Austin's brand receives – if not by direct mentions – subtle support through the preferences of typical music genres. Combined with a focus on organizations mostly related to media and the music business itself, local non-music businesses and alternative artists do not gain the same attention. Live music and the brand is reported on locally but does not get promoted extensively to the general public. It just seems to *be there,* as an integral part of Austin, for now. If media like the Statesmen should decide to shift from reporting mostly on brand-related artists to other genres like latin and indie, the meaning of the phrase *Live Music Capital of the World* might have to be adjusted leading to a stronger incorporation of a broader group of local venders, artists, and urban dwellers. A rather static interpretation of the brand might detach it from the locals even further and finally flush it away with similar memories within the stream of consciousness that are no longer of importance for the local construction of the city.

4.4 Touristic Potentials of Sound

As the previous section has shown, sound can effectively be used to shape a city's image – at least for visitors and foreign tourists. People actually coming to a city like Austin, Texas, often want to learn more about the place where they spend their vacations. Guided city tours are an established way to find out about prominent people, sites, and landmarks. Traditionally, a tour guide leads visitors through her/his urban environment. As a local actor with specific information on the city, s/he presents a certain sense of place to visitors and helps to shape individual experiences. In this chapter, audio-guided tours demonstrate how the traditional guide is replaced by an audio recording wherein the tour's producers to evoke a similar place image use sound.

The first audio guides were used in museum pedagogy. Section 4.4.1 will start with the original reasons for creating audio-guided tours and describe the use of its predecessor, the guided tour led by a human tour guide. Sound will be shown as a way to transport sense of place to multiple listeners without restrictions on day and time. The ubiquitous audio guide narrates a story, using not only the speaker's voice but additional sound categories derived from acoustic ecology. Today's ideas and use of museum audio-guided tours are discussed in interviews with several curators of major museums in Austin, Texas (4.4.2). This will lead to sound composition and the extension of guided tours into the city context (4.4.3). To deconstruct a media soundscape like the one of an audio-guided tour, Schafer's *keynote*, *signal*, and *soundmark* trichotomy (1994) is complemented by three new categories *storyteller*, *soundframe*, and *citation* (the first two deriving from a study on detective novel audio drama, as presented in Chapter 5 (Wissmann and Zimmermann 2010)). A media analysis of all existing audio-guided city tours in Austin, Texas, will reveal how single soundscape elements are used, and how sense of place can be created through audio media.

4.4.1 Concept and Reason of (Audio-)Guided Tours

In 1952, the number of visitors listening to curators of the Sedelijk Museum in Amsterdam dropped not because of a decline in interest in the arts but the introduction of a new concept: audio-guided tours (Tallon). From the early 1950s on, listeners have been able to learn about art without booking a tour guide to deliver the information. The tour has already been pre-recorded. Instead of having the same curator explaining the same topic over and over again – which might result in a less engaged performance over time – audio guides deliver the same facts at the same exhibits. Everything the narrator tells the listener is additional information to that on the exhibit's signs or accompanying text. It even gives visitors the freedom to look at a piece of art without interrupting the visual experience by reading text from a brochure or sign. In comparison to curator-led tours the listener of an audio tour can walk on her/his own. There is no group of people standing in the way, again disrupting the visual experience.

Today, a lot of traditionally human-guided city tours are offered as audio tours, as well. In addition to the audio files that deliver the information about several stopping points, a map is needed to get from one stop to the next. As stated above, a guided tour is often used by visitors who want to learn more about what makes a city such a special place. To tell participants of a tour about a city's sense of place is one of the key functions of the tour guide. S/he usually is a local expert, who has lived in town for a fairly long time and knows about "'free' resources" like the city's history and its unique culture, social life, and the importance of specific places (Wynn 2010: 160). In face-to-face situations, s/he tries to capture the city's characteristics and acts as a cultural mediator between the city and the visitor. The marketing and framework of the tour, commissioned by the local tourist office or another player in the tourism segment, also has to be communicated. Beliefs, ideas, and images have to be translated not only linguistically but culturally (see Dahles 2002: 784, Macdonald 2006: 119).

In addition to knowing the facts about specific stopping points and being able to communicate those to the visitor, the tour guide is also a social actor (Wynn 2010: 148). As a local, s/he represents the city. In the live-tour situation – and this already shows a huge difference in audio-guided tours – the guide has to respond to different types of people, whatever cultural context they come from, how old they might be, or what kind of educational background they have. The tour guide has to adjust her/his explanations every single time. In this "extended customer-contact", social skills are a major quality of the tour guide for reaching as many visitors as possible (Huang, Hsu, and Chan 2010: 4). Depending on group size, age, nationality, and other determining factors, the tour guide ideally delivers an always-different performance for each group, adjusting the stories and comments and even casually making jokes (see Pizam 1999).

Storytelling is a keyword for guided tours. Local history has to be brought to life in the narration of the tour guide, giving the visitor a vivid image of the city. It can be described "as a way in which tour guides explain the urban fabric to others" (Wynn 2005: 400). As visitors understand their own cultural background best, stories are often adjusted and adapted to – phenomenologically speaking – their streams of consciousness. The meaning ascribed to landmarks, local heroes, and traditions (see Kelner and Sanders 2009: 138) may be altered but the key message has to be the same. The city's sense of place – at least the promoted one – has to be sold to evoke a consistent image (Dahles 2002: 784). Thus, Huang et al. call the tour guide a "frontline employee […] in the tourism industry" (2010: 3).

The experience of every listener is shaped by the narration of the tour guide. Depending on the individual history of the listener, a tour is always perceived differently (Cohen 1979: 193). It is because of the communication skills of the tour guide that the perception of the city is not as variable. Storytelling is a dynamic element to evoke sense of place (Wynn 2005: 399) and to help visitors deconstruct the cacophony and bustle of an urban environment (Banerjee and Southworth 1990: 90). It is precisely the dynamics of a traditional guided tour that are missing in the audio pendant of city tours. What might not be a problem in a museum can be a real challenge in the urban context. The pre-recorded narration in an audio file fails to adjust its storytelling to each individual listener. Thus, guided tours become not only less personal, but might even fail to communicate the desired sense of place of a city. But before the problems and the overall potential of those tours are discussed (see 4.4.3), the downsides of traditional tours have to be considered, which led to the implementation of audio-guided tours and their becoming available in almost every major city.

"For dwellers and visitors of any age, a metropolis can be confusing and overwhelming" (Wynn 2005: 400). This not only applies to visual impressions but also the urban soundscape. Following a tour guide is a rather passive act (Haigh and Gold 1993: 22). The voice of the guide might be masked by sounds of traffic, constructions sites, and other sources of noise (Deubel 2007), making it hard or almost impossible to understand what s/he is talking about – particularly when standing in the back of a group of more than 20 people. There is a predefined time schedule that will most likely not suit every visitor (Bradbeer 1996: 12).

By pre-recording the narration, audio-guided tours eliminate many of those downsides of traditional tours (see Wissmann 2013: 1–2). Now, visitors can walk at their own pace and take the tour at their desired day and time. The sound quality of an audio tour is – especially in a noisy urban environment – much better than the traditional listening experience. Whenever the listener cannot follow the narration, s/he can rewind the recording and listen to it again. It is also possible to only visit the places the visitor wants to see since s/he is free from the requirement of following a given route and free from stopping at any points s/he has no interest in.

The idea behind implementing audio-guided tours is to replace the downsides of traditional tours described above with the mentioned features of the audio files. Given the existing conditions, sense of place seems to be much harder to communicate. This is not a big issue for museum audio-guided tours, which generally focus on individual exhibits. As the following examples will show, the focus of these tours lies somewhere else. Offering visitors the choice to choose from different media might be motivation enough to use sound in the museum context.

4.4.2 Idea and Use of (Museum) Audio-Guided Tours – Austin

In Austin, capital of the state of Texas, many museums are open to the public. Visitors have the chance to learn more about paleontology, recent state history, and contemporary art, and take a look at paintings, sculptures, and installations. Overall 17 larger museums are promoted on the city's Web site, and they receive a wide range of visitors, from children (Austin Children's Museum) to students (Lyndon Baines Johnson Library and Museum) and adults (e.g., the French

Legation Museum). Of those 17 museums only six claim to use or have used audio technology in their exhibits (see Table 4.4). Currently, only three venues offer an audio tour (Austin Museum of Art, George Washington Carver Museum and Cultural Center, and the Texas Memorial Museum). The technologies used to implement the tours vary not only in complexity but also cost and sound quality. There are high-end productions with multiple sound effects, several narrators, and pristine sound quality. Those tours have to be scripted in advance, and music and citations from artists have to be pre-recorded or bought.

Another approach is the Guide By Cell audio tour, named after the company providing museums with this service. The curator simply can has to call a phone number to make a recording – as easy as leaving a message on a voice mailbox. Afterwards the museum's visitors can access each recording by dialing a certain phone number to replay the recording. This method is cost-effective, easy to set up, and fast to update, so much so that venues like the Austin Museum of Art and Texas Memorial Museum use (d) this technology.

Table 4.4 Museums in Austin, TX, (Not) Using Audio-Guided Tours

Arthouse
Austin Children's Museum
Austin Museum of Art
Austin Children's Museum Laguna Gloria *
Blanton Museum of Art **
The Bob Bullock Texas State History Museum **
Elisabet Ney Museum
French Legation Museum
George Washington Carver Museum & Cultural Center
Harry Ransom Center
Lyndon Baynes Johnson Library & Museum
Madison Log Cabin
Mexic-Arte Museum
Mexican American Cultural Center
O. Henry Museum
Texas Memorial Museum
Umlauf Sculpture Garden & Museum

Note: Shading: Using (white) / Not using (grey) audio-guided tours.
* No longer exhibiting **Not using audio-guided tours any more.

A large number of audio guides can be downloaded for free from museum Web sites or aggregators like iTunes. They can also be picked up (in a handheld device) at the museum's entry – as in the Blanton Museum of Art – though this is less common. But contrary to these methods, the cell phone guided tour seems to have become largely accepted now that almost everybody owns a mobile device. Exhibit signs provide visitors with a simple phone number to call so they can listen to the related audio content like they would listen to their voicemail messages. "The main purpose is to provide additional information to understand a specific work of art" (Mellard 2011). One downside of this method clearly lies in the often-poor sound quality of the recording, which can get even worse if a particular phone has bad reception. Another problematic issue is the need to own a cell phone to be able to listen to the content: "If you don't have a cell phone [...] you don't get access to it" (Cid 2011).

Discussions with curators and directors of the museums that use (d) audio guides reveal that the reasons behind implementing audio guides vary.

> I think it was more tradition that what a lot of museums have [...] because those museum goers that go to a lot of museums are familiar with the audio tours. [...] [I]t also helps those who haven't been to a museum. [...] It gives them an overall better experience, fulfilling experience (West 2011).

Other representatives understand audio tours as a tool "to provide some additional information. [...] It certainly adds to your content or pointing out things" (Cid 2011). One major reason to implement audio guides is the flexible use of content in media, as David Denney, the director of public programming of the Texas Memorial Museum, explains:

> The main thing we're looking at is flexibility so we develop that content and then we don't tie it to one particular technology but it's actually used in multiple technologies. [...] I think the flexibility is important (2011).

Such rather pragmatic use of generated content doesn't focus on educational purposes. It points to the fact that the implementation of new technologies not only depends on how easy it is for the visitor to use it, but also on well how the existing museum staff can produce audio-guided tours. If planned production is more sophisticated than recording a message on an answering machine, the museum might even have to hire staff to take care of the new media:

> [We will] have a full time person on staff who is focusing on new media, which will include audio guides. One of the things I'm going to have that person do is pay attention to how it's being used (Denney 2011).

Remarks about the overall quality of the experience of audio-guided tours point to their educational value. Visitors should be able to learn something when they come to a museum. Regardless of whether the audio guide is well produced and easy to handle, it is still "the content that counts" (Denney 2011). Great stories and entertaining sound snippets aren't needed to get people to listen to an audio guide.

Relevant and interesting information about an exhibit is far more important. Of course, production quality and a high entertainment factor can't help but enhance listener satisfaction and bring in an even higher number of visitors.

> I think that's a tangential purpose. I think the main purpose is to provide additional information to understand a specific work of art [...] It addresses people who learn in other ways than visual means. [...] It allows you to provide contextual information that wouldn't be apparent just from a visual review of a work of art. It allows different voices into the museum that could literally be the voice of artist who made it. [...] And then a side effect is that someone might spend longer in front of a work of art than they would if they just walked through on their own. [...] Also it's self-directed compared to if you get an audio guide and it has these stops that you have to go through. This is just you can kind of sample, if you want to know more about something you can find out about it you don't have to (Mellard 2011).

Sampling is a quality that most experts describe as being desirable (see Cid 2011, West 2011). But the visitor's freedom to rearrange the order of stopping points and to look at only those exhibits that s/he is most interested in limits the potential of the audio material, which ideally provides the chance to interconnect single stopping points and strengthen the overall stringency of the tour. If there is a connection between several exhibits (e.g., an artist's influences and his/her socio-cultural context) it could be interesting to let the visitor know about it and thus add understanding and depth to single pieces and the exhibition as a whole.

> If you're trying to connect a story together that's very difficult, you can get highlights, you can give experiences. The audio tour fills in that function. It helps create that A, B, C, D experience (Denney 2011).

The interconnection of stops is especially important when it comes to guided tours in the urban environment where a number of locations are chosen to represent the city and to communicate its sense of place – an issue we will discuss in the following section.

Surveys taken by several museums show in what diverse ways audio guides are used (Austin Museum of Art 2010, Blanton Museum of Art 2011, Texas Memorial Museum 2011). The Texas Memorial Museum offers a Guide By Cell tour for their permanent collection. From 2007 to 2011 audio information on 34 prompts was requested over 45,000 times by visitors. Over 10,000 calls per year – equaling around 30 calls per day – show the consistent use of the audio guides since their implementation. But a closer look at the data shows that the service was only fragmentarily used. On average, a listener dialed in to fewer than three prompts – i.e., less than 10 percent. Depending on the year, the average minutes per unique visitor varied from 2:45 to 3:08. The number of times a single prompt was dialed shows an even stronger variability of use. The most popular phone number was called 4,475 times, while the least popular one only received one single call. As phone numbers generally follow the route of the tour – even though some higher numbers have been added by the time – the popularity of single prompts tells us about the popularity of an exhibit but also about the selective use of the audio material. Prompt 1 was called 1,141 times, prompt 2 got 4,475 calls, followed by prompt 3 with 3,367, and prompt 4 with only 961 calls. Afterwards the number went up again to about 1,500 (Texas Memorial Museum 2011).

Similar results concerning individual use of single prompts came from the Austin Museum of Art exhibition Chris Jordan. While prompt 4 was listened to 207 times over a period of 75 days, prompt 5 was way more popular, with 477 calls. Prompts 6 and 7 were significantly lower (187 and 120 calls, respectively), while prompt 8 received only eight calls, and prompt 9 no calls at all (Austin Museum of Art 2010).

The Blanton About Face survey reveals that about two-thirds of the users of audio guides were not new to the museum. The majority visited three to five galleries per year. The ages of the listeners were rather balanced, while there were a lot of more females using audio-guided tours (77.9 percent) than men – and around 80 percent had a four-year college or a Master's degree.

The intention to use an audio guide coincides with the assumptions of the experts: "To learn more information about the art or exhibition" outnumbers answers like "Easy access" (55.7 percent to 8.6 percent) (Blanton Museum of Art 2011). The audio-guided tour "brought a lot of the components of the exhibit alive that you couldn't do walking through just reading labels. It was a lot more compelling, a lot more memorable. It was a better experience. People chose to utilize it" (Denney 2011). "[T]here are [...] people who, they want to know everything that you tell them about it [...]. There are other people who really like technology and they would want it that way" (Mellard 2011).

Although not every visitor to a museum prefers to use audio guides, it seems there is some potential left to optimize existing tours to make them an even better experience. This often-cited "enhanced experience" (Cid 2011, Denney 2011, Mellard 2011, West 2011) could be upgraded by delivering a more diverse production than narrator-only audio guides. Even the Guide By Cell concept allows for the pre-production of audio content that can be uploaded to the voice mailbox system. But no museum – at least in Austin – makes use of this opportunity. As the *About Face* survey shows, visitors wish for more diverse sounds in the recordings. Asked what else they would like to listen to in future tour content, commentaries by *living artists* (67.7 percent) and *art historians and curators* (47.7) received the most answers. *Music relevant to the artwork* (46.2 percent) was named the third most wanted content (Blanton Museum of Art 2011). The variety of elements in an audio tour's soundscape has another, secondary, purpose, which will be discussed in the following section: the creation of a sense of place through sound.

In addition to a richer soundscape, embeddedness can be a way for audio tours to bind the content of single stopping points together. But one exhibit can only be connected to the following if there is a mandatory route and an *A*, *B*, *C*, *D experience* (see above). Most museums don't reference one exhibit to the next "because the assumption is that they are all stand alone. [...] [Also,] the audio guide is usually produced before the art is hanging on the walls and you might not know what's going where [...] That's a practical concern, it's not a philosophical approach" (Mellard 2011). Another argument about not leading listeners through their experience is that "part of the learning experience of viewing art is to see where you enjoy the piece best. I think that's something you have to learn yourself. That may be more intuitive" (West 2011).

The audio tour as additional information always influences how visitors experience a work of art. While exhibits might be placed in a more practical way, the urban environment consists of more purposeful patterns. Questions about why a certain building has been erected close to a park, highway, or within an area like the central business district can be explained through an audio tour. While museums largely decide to not integrate relational information in their audio guides, city guided audio tours could be using this method more distinctively to tell the story of a city's sense of place. Even if the following section shows that they don't do this, the idea behind and the use of audio tours stays the same: Whether listeners are "streakers, strollers [...] [or] studiers [...] [they] have their own ideas how to experience" their environment (Denney 2011). Audio media is an attempt to enrich a visit and satisfy every listener by adapting the given information to her/his personal habits (also see Chapter 5).

4.4.3 Sound and Effect of (City) Audio-Guided Tours – Austin

Audio-Guided city tours picked up the concept of museum tours established in 1952 in Amsterdam's Sedelijk Museum. In this medium, the tour guide as mediator of a city's sense of place who individually tells visitors about urban places of interest and the city as a whole is missing. As discussed above, producing audio information for single stopping points that are not interconnected might make it difficult or even impossible to transport and create the city's special meaning. Sense of place, defined as "center of felt value" (Tuan 1974: 138), that bares social meaning (Halpenny 2010) and illustrates the dominant image or theme of what the city *is* (Cresswell 1996) does not come from the urban environment itself without guidance. As listeners of an audio-guided tour have to rely on their experience and perception combined with the additional audio information, the latter has to carry all the meaning of the city – at least the one marketed by the local tourist bureau.

In a city like Austin, Texas, there are several audio-guided tours that discuss diverse urban topics, produced by local interest groups, independent artists, and officials like the state or city. In Austin, 11 tours can be found on the producer's Web sites, audio aggregators like iTunes, or at the place where a tour actually starts. More than half of the Austin tours are art installations promoting the Domain, a suburban estate combining living, shopping, food, and art in one place. To describe the concept behind the Domain six audio recordings were produced by local artists like Alex Keller (2009), Patrick Robinson (2009), and Vanessa Rossetto (2009).

Table 4.5 Austin, TX – Audio-Guided Tours (Including The Domain)

A Audio-guided tours in Austin, TX (2012)				
Name of the tour	**Producer**	**Field of interest**	**Tracks**	**Soundscape elements**
Audio Tour for Works of Public Art	Landmarks	Public art	18	Storyteller, soundframe, citation
Audio Tour of the Lady Bird Johnson Wildflower Center	Lady Bird Johnson Wildflower Center	Floral exhibition	20	Storyteller
Austin Walking Tour	Texas State Travel Guide	City guide	11	Storyteller, soundframe, soundmark, keynote
Barton Springs	Save our Springs Alliance	Environmental issues	13	Storyteller
Historic Walking Tour Texas State Cemetery	Texas State Cemetery	Cemetary history	37	Storyteller
B Audio recordings for The Domain, Austin, TX – Art installations "Recreating the Domain"				
Name of the tour	**Producer**	**Field of interest**	**Tracks**	**Soundscape elements**
Border's card	Patrick Robinson	Art installation	1	Storyteller, keynote
Century oaks	Vanessa Rossetto	Art installation	1	Keynote
Curator's notes	Alex Keller	Art installation	1	Storyteller, keynote
Dominion: a walking tour	Brent Fariss	Art installation	1	Storyteller, keynote
Recreating the domain	Chrstopher Petkus	Art installation	1	Storyteller, soundframe
Tan bodies	Bill Bridges	Art installation	1	Storyteller, keynote

These audio-enhanced urban experiences directly communicate with the listener, sometimes without even using a narrator but only keynote sounds (see Table 4.5). They might be called 'audio tours' only in a broader sense because they don't use predefined stopping points and additional material like a map to help the listener navigate through a part of the city. Walking instructions are limited, as one-track-only tours work with the interplay of acoustic stimuli and visual environmental impressions. Thus listeners might learn more about themselves – they are sometimes even asked to perform certain actions – than landmarks and other places of interest.

Traditional ideas about and uses of an audio tour are far more apparent in the other five audio tours that are located all over the city of Austin. As Table 4.5 shows, the fields of interest of these tours range from cemetery history (Historic Walking Tour Texas State Cemetery) and environmental issues (Barton Springs), to public art (Audio Tour for Works of Public Art) and local flora (Audio Tour of the Lady Bird Johnson Wildflower Center), or they act as a replacement for a traditional city guide (Austin Walking Tour). They all contain multiple stopping points and walking instructions, can be downloaded for free off the Internet, and can be used on any audio player like cell phones or iPods.

The various audio files differ not only thematically; audio-guided city tours also vary in production quality and soundscape elements. Soundscape elements have been discussed when acoustic ecology's approach to deconstructing the environmental soundscape has been presented (see section 3.1.2). Keynote, signal, and soundmark can also be found in media soundscapes, where their sounds signify the same things they do in an urban environment. The keynote describes all background sounds in the recording, like traffic noise, wind blowing, or undefined murmurs of multiple conversations. Signals stand out of the keynote, like the bark of a dog and the siren of an ambulance. Soundmarks in an audio recording can refer to the surroundings in which it is listened to – like the ringing of bells in a church next to the listener – but can also stand for a place-based sound inside the recording, as Wissmann and Zimmermann show with audio drama: "The headquarters located at the salvage yard, which is used frequently throughout the entire audioplay series, is acoustically marked by the soundmark of a parrot's squawk" (2010: 377) (see 5.2).

But "[t]he given three-way breakdown of a soundscape is insufficient to cover the events of an audio-play. [...] In addition to keynote, signal and soundmark [...] there are three additional distinct elements: storyteller, dialogue and soundframe" (Wissmann and Zimmermann 2010: 376), two of which can also be found in the audio-guided tour. The *storyteller*, as narrator of the story in audio drama, signifies the voice of the curator or tour guide, while the *soundframe* represents mostly musical sound elements, like jingles to announce a change of scenes within audio drama or the arrival of stopping points within an audio tour. As a sixth soundscape element, the *dialogue* of characters in audio drama is replaced by *citation* – citing of artists, experts, or excerpts from other media like film within the audio tour.

As the listing in Table 4.5 illustrates, some soundscape elements occur in only one of the 11 analyzed audio tours (*citation* and *soundmark*), and one element (*signal*) is completely missing. *Citation* is supposed to allow different voices, ideas, and arguments into the tour, as the example of the Landmarks audio tour shows. At stopping point 13, *Clock*, not only Linda Henderson, professor of art and history at the University of Texas, is asked to speak about a sculpture by Mark di Suvero's, but the artist himself adds to the conversation:

> I struggle for … to make a work that is joyous with life, that aids people to be together, that can move people, not just emotionally, but physiologically. So far I've had beautiful responses and I don't know what will happen (Landmarks).

Like *soundmarks* and *signals*, *keynote* sounds are nonverbal elements. The latter can be heard in six tours – five of which are art installations of the Domain – which is double the number of *soundframes*, which are also nonverbal (integrated three times). The one soundscape element to be found in almost all of Austin's audio tours (10 times) is the storyteller. To transport information it is first and foremost language that is used by the tours' producers. Three of the five Austin audio tours outside the Domain use only a storyteller in their productions. One explanation might be that the surrounding environment generally provides background sounds where a person is listening to the recording. Natural signals and soundmarks also occur autonomously. To reduce production cost, citations and soundframes are spared, and the narrator's voice is the only sound to listen to.

> Welcome to the courtyard. This is the heart of the Lady Bird Johnson Wildflower Center. In the middle of the courtyard is our spring. It simulates the natural springs found throughout the Texas Hill Country (Lady Bird Johnson Wildflower Center 2009).

This way important information about a stopping point can easily be provided. The listener can even be addressed directly: "Before the damn was build Barton Creek flowed into an over the main springs of the pool. Is there water flowing in the creek upstream? If not, this means the aquifer is low" (Save our Springs Alliance 2010). If the narration does not set a location in context with other stopping points and its surroundings, additional locations can be integrated easily to the route. Thus, the track numbers of the audio files – or numbers to dial if it is a cell phone guided tour – might not be in order: "Number 55. Introduction. Welcome to the Texas State Cemetery" (Texas State Cemetery).

The minimalistic approach of only using the storyteller in the audio recording makes sense not only for low budget productions. If other soundscape elements are included, they have to add to the purpose of the tour and not distract the listener. If the recording is listened to in the urban environment it does not seem to make sense to add another mix of keynote sounds. Sound of cars passing by mixed up with sounds of actual on-site traffic seems only to confuse the listener. Signals of sounds only to be heard at specific times a day or year might be useful to illustrate the story, though. If the church's bells only ring every Sunday, or natural phenomena like Austin's bat colony under Congress Avenue Bridge can

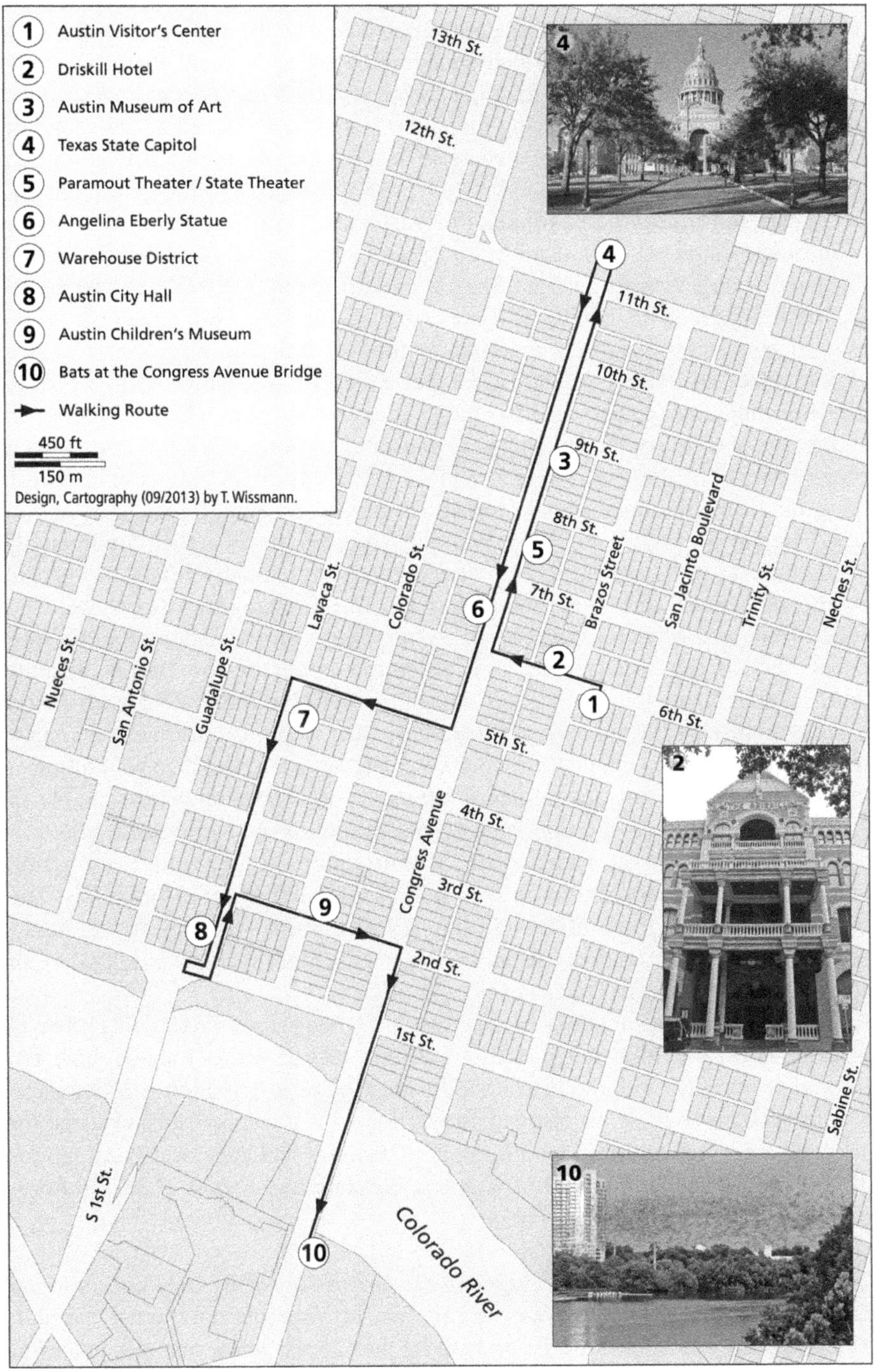

Figure 4.5 Austin Walking Tour

only be seen and heard when the sun goes down in the warmer months (see Austin Walking Tour), signals and soundmarks are reasonable for production. But the recorded sound of an announcement at a train station that gives notice of a fictitious train arriving during rush hour is almost as absurd as the sound of walking shoes that simulate the listener's own walking from one stopping point to the next.

Except for *citations* and *signals* the Austin Walking Tour contains all other soundscape elements. Produced by the Texas State Travel Agency (2012) it is designed to tell visitors to Austin more about the city – that is, to propagate a certain image of the capital of Texas (see Figure 4.5). This image has to be evoked almost exclusively via sound, backed by only the visual impressions of the urban environment a person experiences while taking the audio tour. The environmental soundscape has to be taken into account when producing an audio tour. Because of the existing keynote sounds and signals – sometimes even soundmarks – of the environment, audio-guided city tours almost exclusively focus on the media soundscape elements of storyteller, soundframe, and citation (see Chapter 5). Only the Austin Walking Tour integrates a(n environmental) soundmark at one stopping point. The keynote sporadically refers to historic events or scenes and does not play back the contemporary keynote. The soundframe could be used to refer to a city's sense of place, but no audio tour makes use of the opportunity. One exception might be the Barton Springs tour, in which the sound of water splashing signifies the end of each audio file – symbolizing the audio tour's main topic: the water of Barton Creek. Apart from that, linguistic elements of the soundscape – storyteller and citation – are the most direct way to influence listeners' perception and lead people in a certain direction. As citation only occurs in one of the audio guides, and in principle takes over the function of the storyteller – i.e., giving information to the listener – a closer look at the latter yields interesting information. The storyteller appears in all five audio-guided city tours – and in 10 of the 11 total tours analyzed. Its deconstruction reveals the most variable potential of all soundscape elements to create sense of place.

A narrow context analysis of the Austin Walking Tour (Texas State Travel Agency 2012, all following quotations belong to this source) shows how the storyteller interacts with the listener and what kind of information it delivers. To deconstruct this soundscape element, the audio tour was transcribed and tested for different meanings of the given information. It shows that there are six categories of text the storyteller consists of: *General information, Frame, Stop, Add-ons, Action,* and *Place* (see Figure 4.6).

General information is kind of a special category as it usually only occurs in the first, introductory audio file, where the listener learns about how to use the tour. The purpose of general information is to let the listener know about the concept of the tour and about how to use the audio files and possible additional material

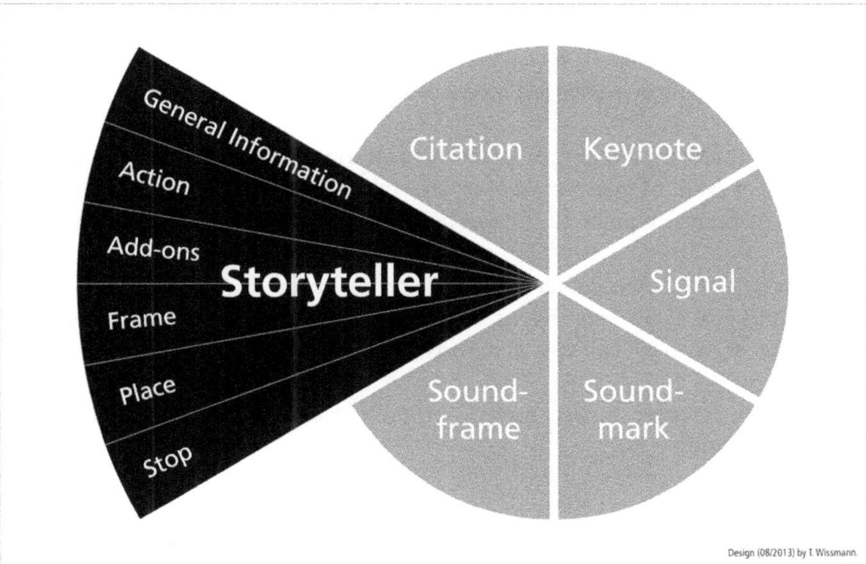

Figure 4.6 Audio-Guided Tours – Soundscape Deconstruction

like a map. "How the tour works: Follow the MP3 player for directions. The map will guide your journey, with each stop corresponding to the numbered segment on the MP3 player" (Stop 1). Safety issues are also addressed, like sensitizing for change of environmental perception because of the audio input. Warning signals and traffic might be harder to perceive.

> Because the digital audio content provided with this tour can block out nearby sounds, please be aware of your surroundings at all times to ensure maximum safety. The tour will guide you across streets and busy intersections so it is important that you remain actively aware (Stop 1).

Frame, as the second of the storyteller's categories, usually occurs at the beginning or end of an audio file to indicate the move to the next stopping point. Walking instructions, often accompanied by a musical soundframe, pause the narration and directly address the listener.

> Cross Congress at the light. Then cross Cesar Chavez. Proceed south across the bridge to the steps. 'Estimated Time to Next Stop: 7:30 minutes' (Stop 10).

Information about the stopping point is summarized in *stop*. The storyteller only refers to the point of interest itself, communicating facts like the date and time of erection, characteristics of the facade, or the symbolic meaning of a landmark.

> The Driskill Hotel is a truly historic Austin landmark; a cherished old building in the heart of downtown, lovingly maintained and still successfully fulfilling its original purpose as a grand and luxurious hotel (Stop 2).

To learn more about a stopping point than just information on the particular location, landmark, or point of interest, *add-ons* are embedded in the text, as the following example for stopping point 10, Bats at the Congress Avenue Bridge, shows:

> Largely, it must be said, due to educational efforts by the Bat Conservation International (who even moved there headquarters here from Milwaukee), Austin began to fall in love with the bats (Stop 10).

The add-ons category is also used to promote a city's brand by talking about touristic highlights:

> Every March, some 15,000 musicians, fans, managers, agents, record company executives, producers, DJ's, journalists etc., all flock to Austin for the annual South-By-South-West music conference and festival. [...] Yes, few, if any, would disagree with Austin's title as Live Music Capital of the World (Stop 1).

To embed everything seen into a larger context, add-ons might refer to more general information on a topic or relevant historic facts:

> By the early 1970's, however, the twin impact of TV and a nationwide drift toward suburban movie theaters, threatened the future of the Paramount (Stop 3).

Action describes all passages in the storyteller's text that try to activate the listener. An engaged explorer of the urban environment should replace the passive listener. To learn more about a stopping point it might be useful to reposition oneself:

> For viewing purposes, the best place to position yourself is down on the south bank of the lake, on the east of the bridge. Get there just before the sun starts to go down (Stop 10).

The action element can also allow the listener to break out of the given stiffness of the predefined route and explore parts of the city her/himself.

> From your current vantage point outside the casually bohemian Halcyon Coffee Shop, you will see that this area is home to a wide range of pubs, bars and restaurants. It is worth wandering around and seeing what takes your fancy (Stop 7).

Action can also influence the tempo of the tour:

> Take a few minutes to browse the landscaping and architecture (Stop 8).

Place, the sixth storyteller category, widens the perspective of the listener by embedding the stopping point into its concrete surroundings and the overall city. To help the listener understand why a certain building was erected in a particular spot, or why a single quarter has a special meaning for the whole city, *place* turns away from narrow description, as the example of Town Lake shows:

> Town Lake is a popular hang out for locals and visitors alike. Actually a part of the Colorado River (that bisects downtown and is marked by a dam at each end), it boasts a ten mile hike and bike trail that attracts all manner of walkers, cyclists and joggers ... (Stop 10).

The stopping point is not the lake itself but Congress Avenue Bridge, which spans the water. Two stops earlier the listener learns about Austin City Hall and its connection to the surroundings: "The waterfall on the south plaza is fed from condensation from the air conditioning" (Stop 8). In the *place* category, the storyteller has the opportunity to interlink stopping points and establish links from one location to other part in town: "Just a couple of blocks west of the Warehouse District (south west of Republic Square) are two more significant Austin music venues" (Stop 7). Perhaps the most important use of *place* is the connection of a stopping point to the city. The chance to have a single location tell the visitor about what makes the city special is the most effective way to communicate an urban sense of place:

> Congress Avenue today is known as the main street of Texas. Like Austin and like Texas, it grows and it changes, but still it effects a balance between progress and tradition, history and change. Old and new sit side by side, in concord and in harmony. The street holds a million stories and hides a million secrets ... (Stop 6).

Analysis shows that *place* is the storyteller element suited best to tell visitors about a city's sense of place. The total runtime of *place* naturally is limited because it is only one out of six potential soundscape categories for an audio-guided tour. But as only *place* offers the chance to transcend information on a stopping point to tell a city's story, one could expect a good part of the storyteller's time to focus on *place*. This is not the case. In the Austin Walking Tour the most dominant storyteller categories are *stop* (46 percent) and *add-on* information (34 percent). Even *frame* (10 percent) and *action* (6 percent) take up more time than *place*. Only 4 percent of the storyteller's activity is classified as emphasizing and fostering the city's sense of place. Thus, audio-guided tours make use of its potential insufficiently. Stopping points are not interlinked, as they are in a museum. What works for the latter has consequences for the experience of the urban environment. If listeners of an audio-guided tour visit one stopping point after another and learn only about single locations – including additional information that are also not related to the city – audio-guided tours do not provide a *city experience*. They fail to serve as mediator between the listener and the city. With the guide as social actor missing, audio-guided tours have to make use of the narrator's voice – i.e., the storyteller – to compensate for this loss. But the lack of place-related information results in an only weak representation of the city and its sense of place. Without a focus on place, audio-guided tours cannot tell listeners about the city but only about single stopping points. It seems that audio-guided city tours today, while leading from one location to the next, fail to deliver an experience of the urban environment as a whole. A city's sense of place is neither resurrected nor created.

4.5 Sacred Silence

How place is being perceived largely depends on individual perception. That goes for sound, too. Possible problems that occur during an audio-guided city tour because of the always-different understanding of the prerecorded audio material might be less obvious. But what about places that co-define themselves due to their sounds and the special meaning that is inscribed into their acoustics? A church, especially the sanctuary, is such a place, one that is socially defined as being holy. Its visual impressions from stained-glass windows, its unique smell of incense, and also its typical sounds from liturgical music, common prayer, and church bells create an environment of sacredness. What is so special about the acoustics of a sanctuary? As in any other place in the urban environment, it is never completely silent in the sanctuary of a church. The acoustics of the hall-like building feature a variety of sounds like the creaking of wooden benches inside and cushioned traffic noise from the outside. During services, believers whisper and rustling cloth, the voices of the choir, and the music of the organ add to the soundscape. What decibel measurements fail to display is a sacred silence that can be experienced in this cacophonic acoustic environment.

Section 4.5.1 will describe findings from a case study of downtown churches in Austin, Texas. It will show the way sound is used in church and how the sounds of the service create an environment for silence. Qualitative interviews with priests, pastors, and directors of the choir define acoustic perception as highly individual and emotional. They describe the silence in the sanctuary as important for worship and essential for coming to know God. In contrast to the acoustic environment of a concert or city hall, silence is a *planned* state wherein something happens that only the individual can process through her/his own senses. The experience of sacredness seems to be related to place, particularly a place like the sanctuary. It is accessible to believers as well as visitors who acknowledge the specific sense of place of the sanctuary.

Sacred silence that adds to a sense of place will be investigated in section 4.5.2. Like the phenomenological alien, which is originary inaccessible, sacred silence is understood as a presence that can only be experienced in its own absence. Based on the phenomenology of Edmund Husserl (1960) and Bernhard Waldenfels (2011), sacred silence in the sanctuary will be discussed as a matter of individual perception and subjective emotion, both of which are embedded in social context. The individual's stream of consciousness builds the framework for the perception of silence and its emotional description as being sacred.

4.5.1 Sounds of the Sanctuary

The church and its music have long been topics of scientific investigations. From medieval church music in English towns (Burgess and Wathey 2000) to contemporary impressions and effects of sacredness on the individual (Pocock 1996), many authors focus on the atmosphere of sanctuaries and their acoustics. The sanctuary is socially constructed as a holy place where certain rules apply, starting with regulations on how visitors and believers are supposed to behave when entering it.

In contrast to ancient churches in European cities like Paris, London, and Rome, the houses of God in Austin, Texas, are not open to the public during the day. At night, the sanctuary is closed and does not allow just anyone to walk inside, as all of the interviewees confirmed (Barr 2012, Edge 2011, McDonell 2011, Stevens 2011). Although, this was not always the case:

> No, we tried that and, about 20 years ago, said this church should be really open particularly around noon, particularly for people that work in the state buildings around here to come and pray (Heenan 2011).

Because of incidents, mainly with homeless people, church officials decided to keep doors closed except for services and individual enquiry, or special occasions like music festivals (Barr 2012, Edge 2011, Stevens 2011). In the context of Austin's brand as the *Live Music Capital of the World*, some churches like Central Presbyterian Church temporarily become music venues for festivals like SXSW.

> Last year we had a little over 7,500 people in those days come in and sit in our sanctuary and listen to music. Some of it is *rock 'n roll*. Some of it is *alternative*. Some of it is *gospel* (McDonell 2011).

St. David's Episcopal Church even rents out its sanctuary to Buddhist, Hindu, and secular groups to "make quite a bit of money" (Stevens 2011).

Aside from special occasions, when a rock guitar replaces the organ or secular discussions take over from religious preaching, the behavior of people entering the sanctuary seems to be predefined, as "people are very quiet when they come in" (Edge 2011). Observation shows that visitors take a walk around the building, sit down and "pray, light a candle, just be quiet for a while" (Barr 2012). As there are no explicit rules or signs that state the importance of silence, why do people in a sanctuary avoid producing sound? It seems interesting that silence is assumed in "a place of prayer" (Barr 2012). Silence as a "natural response to entering that space" is simply explained as a "tradition" (Edge 2011). But the special acoustic behavior of visitors might also arise out of "a sense of reverence and awe and the presence of God" (McDonell 2011). Sound in the sanctuary is – at least in the context of religion and sacredness – something people seem to have a *sense* or *feeling* for that transcends pure education and socialization. It is an important element for place identity, not only deriving from religious beliefs but otherness in relation to similar large buildings with corresponding acoustics.

> Good acoustics of Orthodox churches typically have high ceilings, a certain amount of reverberation. [...] It's great for chanting and horrible for preaching (Barr 2012).

While the height, width, and length of a sanctuary might impress visually, acoustically they are a challenge for every service. Often, the ratios are wrong, leading to massive reverberation, making sound, especially voices, extremely hard to understand. As Cavenaugh et al. point out:

> [W]orship space acoustics consists of two parts – speech and music – that are acoustically contradictory: a natural acoustic environment that provides high speech intelligibility is very different from one providing good acoustics for music (2010: 151).

But acoustic downsides like the cancellation of soundwave motion and reverberation lead to an experience that co-defines the sacredness of space: "I don't know anything about acoustics but I know this is alive. This sanctuary is alive with sound" (Heenan 2011). Such effects emphasize the music that is composed for these spaces. It sounds different to performances in other surroundings – this may be one reason why event managers turn sanctuaries, as in the case of some Austin churches, into live-music venues. In the case of liturgical music, "reverberant spaces" are not only a nice sound effect but are "fundamentally important" (Cavanaugh, Tocci, and Wilkes 2010: 151). It is an acoustic help for believers to experience the sacredness of place:

> I feel like the music that we make is a doorway to let people walk through out of whatever mundane that they're in. They're worried about their grandchildren when they come to church. They're worried about paying the bills in the afternoon. But the stained glass windows or the incense or the clothing or whatever I think allows people if they will that they can walk through a door and be in the Holy. They acknowledge that it's Holy (Stevens 2011).

Sound in a church, i.e., a sanctuary, seems to have not only an acoustic quality but an emotional one, too. It is described "as a means to guide us" (Barr 2012), which implies that not only notation, tone, and tempo matter. Taking aside the meaning the listener ascribes to it – which from a radical constructivist perspective is impossible and makes no sense phenomenologically – sound appears rather unromantic, even in a holy place like a sanctuary:

> All it is, is the air molecules being disturbed and in the same manner they disturb your eardrums. Then somehow we translate those electronically to some idea or perception [...]. That makes us feel something. All it is, is sound waves. All it is, is air molecules going ba, ba, ba, ba, ba. It's neat that such romance, such depth of feeling, so much information is given and received by sound (Stevens 2011).

The human voice is one of the acoustic stimuli in a sanctuary that can evoke such deep feelings. In the Orthodox Church, for example, "we say that the highest sound we could offer God is our own voice" (Barr 2012). Believers express their devotion through chanting and common prayer. The voices of people talking in the hallway or even in front of the sanctuary are also described as typical sounds of a church. The sound of people's voices expresses a "sense of community" (McDonell 2011). The rustling of cloth, clang of the offertory plates, and creaking of wooden benches all add to the soundscape of the church, while the singing seems to be rather unique. "I don't know any other place, a public place, that you hear singing [like this]" (Stevens 2011).

Musical instruments are also typical for the church's soundscape. Mainly organ and piano represent liturgical music, and are most often associated with a sanctuary. The organ, in particular, is noted as the traditional instrument for service and worship, though over time smaller sized pianos or keyboards have often replaced it.

> We're getting more and more folks coming back that miss that traditional sound of the organ. They miss that. They want that. These are young people telling us that (McDonell 2011).

Asking interviewees about the most typical sound for a church one element that was never named was bells. To the surprise of the interviewer, church bells were valued as important but not as the unique feature of the house of God. From a European perspective this would seem to be impossible as almost no church in countries like Germany comes without church bells. This is different in Austin. Only one research site (St. David) has a bell, which was not even designed for a church but the State Capitol. In principle, the effect of the church bell is acknowledged to "identify the place" of and around the church as Christian territory (Edge 2011). This immense meaning has led to the prohibition of ringing the bells in countries like Soviet Russia. The bell as a symbol for "a traditional way of life sustained by religion" runs contrary to socialist beliefs and therefore had to remain silent (Hernandez 2004: 1478). It was understood to "reinforce social cohesion" and serve "as the authoritative voice" of an antagonistic system that might undermine the political and social status quo (Hernandez 2004: 1478). Another secular but way more harmless perspective on church bells is the telling of the time. People are acoustically informed of a soon-to-start service, a wedding ceremony, or a special day in the calendar of saints. In addition to being informative, church bells also express the sacredness and holiness of place.

> The church bell [...] is an apotrapaic [*sic*!] instrument, intended to sanctify a holy place of holy time. It is centrifugal in the sense that it frightens off evil spirits, and centripetal in the sense that it draws people together for collective religious observance (Bull and Back 2003: 26).

In addition to the sanctifying character of bells, they are also "giving witness to the world of, hey remember? This is God. This is about God" (Edge 2011). This use of the sound of bells can also be found in the Holy Bible, wherein bells are tolled and trumpets are blown to tell of the existence of God (Numbers 10:10, King James Version). In this sense, bells can be understood as a primary "aural icon" of Christianity that brings "eschatological reality" and the awareness of an all-powerful God into being (Hernandez 2004: 1481) and "inaugurate[s] the sacred in time" (Nelson 2007: 236).

Bells, the architecture of the sanctuary, and the human voice are three major elements of the church's soundscape. "There is also that sense of silence is important, too" (Barr 2012). Like music, silence sounds different in a sanctuary and adds to the experience of sacredness. "Quietness is what you need to not be distracted" and focus on prayer and worship (Stevens 2011). The statement "I think one thing there is a, even if it's momentary, there's silence that precedes or interludes or remains when everybody is gone" (McDonell 2011) transcends pure acoustic perception. The absence of sound seems to be inherent to place experience, and its significance has grown in urban environments that are constantly getting louder and noisier.

In acoustic ecology's standard book, *The Soundscape*, Schafer does not understand:

> the negative way in which silence is described my modern writers [...]: solemn, oppressive, deathlike, numb, weird, awful, gloomy, brooding, eternal, painful, lonely, heavy, despairing, stark, suspenseful, aching, alarming. The silence evoked by these words is rarely positive. It is not the silence of contentment or fulfillment (1994: 146).

It may be because of such negative connotations or, more likely, the constant rush and hurry in a city these days, that people are not taking the time to take a break and enjoy silence.

> We're a noise filled society I think. [...] So we try to be busy all the time and I think that's counterproductive to joy and happiness [...] So I think this whole

> sonic space you create for yourself [...] has to include silence. [...] When we don't we're in trouble (McDonell 2011).

As discussed in section 2.2 sound can cause pain and have several primary and secondary negative effects on people's health (Byrne, Michael, and Tufts 2011, Kluizenaar et al. 2011, Stansfeld and Matheson 2003). So "we need that silence [...] to recharge" (Barr 2012). This relaxing and pleasant kind of silence within a city is usually associated with urban green spaces, where natural sounds mask most disturbing noise (see 2.4, especially Mylott and Dubois Williams 2007, Pheasant et al. 2008). The silence of a sanctuary can have a similar effect. It seems like, even without a chance to experience absolute silence acoustically, it can be perceived within sacred space. Even among all the different sounds forming the soundscape of a sanctuary, there is always silence, e.g, in between moments of sound. Together with sound, silence creates a sacred atmosphere, like in (liturgical) music. John Cage's piece *4'33"* has been mentioned before (Kania 2010), expressing that "silence is part of music. When you hear great symphonies they use silence. That's part of the music. That's part of the effect" (Heenan 2011). On the one hand, silence helps to strengthen the impact of sound: "Where there's no sound it sort of amplifies the sound when there's sound" (McDonell 2011). On the other hand, especially in environments like a sanctuary it seems to mask all other sounds – not because it is louder, but because of the intensity of how it is perceived individually. Answering Schafer's statement that there "are no earlids" (Schafer 1994: 11), "it seems to me we do have ear lids. Even if something is hitting our eardrums, it may be being sent somewhere that there is no perception" (Stevens 2011).

Thus, silence is crucial for the construction of sacred space. The reason for this might be that the majority of people entering a sanctuary keep their voices down because of an anxiety related to silence. "Maybe we don't want to be distracted with noise from what we are sensing with our heart, with our eyes, with our minds when we enter that kind of space" (Edge 2011). Personal, inner silence helps people get into the sacredness of place. Inner tranquility helps mask surrounding unwanted sounds, like outside traffic noise. Even if visitors to a sanctuary talk in low voices, cloth rustles, and wooden benches creak, "In those big acoustics I think sometimes you do get a sense of aloneness even when there's noise" (Stevens 2011).

Another reason to be silent – aside from the personal need to flee urban cacophony – is cultural norms. "One major characteristic of most Western cultural silence [...] seems to relate to conceptions of man-made authority and conceptions of highest authority or God" (Bruneau 1973: 37). We may be in awe of God or a church's architecture or unique acoustics: Visiting a sanctuary requires some kind of "church voice" (Barr 2012), a certain way of behaving while talking, praying, and singing that is different from the normal use of our voice. In Christian ritual silence is a controlled silence (Bruneau 1973: 38). During service there is "the need to be silent" that comes with the experience of the sacred (Edge 2011). "Holy space induces stillness, quiet, and invites us to linger" (Pocock 1996: 382). It seems that the Greek orthodox practice of hesychasm [ἡσυχασμός], the prayer of silence, almost replicates "the silence of centuries" (Pocock 1996: 384) that lives in the church as a social construct.

Being silent inside the sanctuary, especially during a service, gives us "the opportunity to worship in a very unique way" (Edge 2011). Sounds are experienced against a background of silence wherein even the most quiet creaking in the wooden rafters is perceived and ascribed some kind of sacred meaning: "We always say that's all the saints that have gone before us that are sitting up there in the rafters" (McDonell 2011). This quote may not be taken too seriously but it shows how sound is reflected on in places like a sanctuary. Because silence emphasizes the sacred experience (Edge 2011), it is built "into our worship service" (McDonell 2011). In between different elements of a service like the choir singing and the pastor preaching, silence helps to differentiate and emphasize. It is also important to create a non-daily atmosphere where believers can focus on their faith and conversation with God. "Here we have silence a few seconds, 15–20 seconds, before a prayer, before the pastoral prayer. I think that prepares people for that prayer" (Heenan 2011). Without silence it might be too hard to concentrate on prayer and worship. Noise could easily distract people and prevent them from diving into the experience of the sacred. Acoustic elements like communion music are often performed in a non-disturbing way.

> I don't want to say easy listening, but something that won't be too distracting. Hopefully it adds to their worship experience. That's what it all is, a worship experience to me (Heenan 2011).

The individual choice and possible "need to be able to sit in silence" (Edge 2011) can be completed with an active refusal to commonly sing or speak out loud during a service. It does not have to be a rejection of certain rituals and a missing sense of community that causes people to decide to stay silent. "From the Pastor's point of view you look out and see many people don't sing. They don't open their mouth" (Edge 2011). It might also be an individual, internal continuation of the soundscape inside a sanctuary that partly gets its sacred effect because of silence. The sacred meaning of places like the church and its reduced sound can have the effect of masking other, external, unwanted sounds.

> But when I go in there during the day to sit and pray, reflect, if I hear the sounds of traffic from the outside it doesn't keep me from doing that which I'm there to do. I might notice it but it doesn't bother me (Edge 2011).

The atmospheric context lets otherwise disturbing sounds appear less important. What is it that makes a place holy or sacred? The holiness of the church is hard to grasp, as the previous discussion shows. From a secular point of view it seems to have something to do with the overall visual, olfactory, and auditory impressions people get when entering a building known as a house of God.

> It doesn't really look like their school and it doesn't really look like their living room. [...] Of course that's part of what music and stained glass, incense, and all that kind of stuff does. As far as I'm concerned [...] it's holy here (Stevens 2011).

The importance of sound to the holiness of place seems to arise out of its being different from the usual urban soundscape. This does not mean that in a sanctuary there are completely different sound events occurring, but that common sounds are experienced differently.

> Whatever sounds they're normally hearing throughout the week, they hear some of it at church but there's something different about that sonic space as opposed to one's home, workplace, or neighborhood (McDonell 2011).

Being different from day-to-day life make the church something extraordinary, something that is, in a way, *out of order* and "not of this world" (see next section) (Barr 2012). The sacredness of silence seems to be located in this rather inaccessible place, which believers can never fully grasp but constantly try to reach out for. But sacredness also seems to be something that is not only to be experienced by believers but almost anyone entering a place like a sanctuary. Religious affiliation does not play a role in acknowledging the holiness of place.

> If we went into a mosque, if we went to Israel and Jerusalem to the Dome of the Rock it may be something I want to go see as a tourist I'd still think this is a place of prayer (Barr 2012).

Even for non-religious people sacredness might be sensed. Atheists and nonbelievers do not base their perceptions on the same (religious) stream of consciousness that believers do, but entering a church almost always evokes a sense of respect and awe, or as one of the interviewees put it: "You might not be in love with my wife but if she was the most beautiful woman in the world you'd feel something when you saw her" (Stevens 2011).

4.5.2 Experiencing Sacredness

Silence, as part of the sanctuary's soundscape, plays a role in evoking sacredness. But how exactly can the absence of sound help to enable believers or even visitors to experience the holy? Descriptions about what silence and sound in general do in the church are rather vague, as the following example shows: "To me what's so beautiful about a great acoustic is you have a visceral experience" (Stevens 2011). Emotions are a popular way to describe it. Actions performed to achieve a state where sacredness can be felt are rationalized by a pre-discursive consciousness that shows itself in rituals. "A church family has rituals, too, that are brought about by the use of sound" (McDonell 2011). Generally, the acoustics of a sanctuary are recognized as creating an "exalted mood" (Blesser and Salter 2007: 2). The building itself is ascribed a *soul* that comes to live in sound. The music of a concert inside a cathedral transforms the stones and evokes memories of the past (Ihde 2007: 50–51) and moves listeners beyond measure.

> There was like 7 and 9 second reverb and it was kind of a slow reverb and they were singing some polyphonic piece up there. You couldn't even hear what it was. All you heard was the blur of lines. Now, that's almost like seeing stained glass windows against something and seeing the blur of color. I think that I have really felt incredibly moved sometime by hearing clustery music like that. Even if it wasn't supposed to be clustery but it was very beautiful but it's because it surrounded me (Stevens 2011).

Sound – in this case its organized form, music – has something extraordinary that moves the listener. Stevens tries to explain the emotional effects of sound by comparing an auditory experience to a visual perception that touched him. Silence as part of the soundscape also reflects the tradition, history, and meaning of place. It is a "mute testimony of the past" and can be experienced in rituals like silent prayer (Ihde 2007: 51). Privacy is especially supported by silence. The personal dialogue with God, the "privacy of quiet contemplation", is enhanced whenever no unwanted sound disturbs the moment (Blesser and Salter 2007: 2).

> In worship we talk about how silent moments are precious. Just like at the end of the choir's presentation on Christmas Eve there was that pregnant pause, that sacred moment to take a deep breath and you soak it in before you go on [...] (Edge 2011).

The silence right before the next prayer is a mystical moment that prepares believers to approach God (Zembylas and Michaelides 2004: 197). The individual decision to keep quiet and silently communicate with God is variable. Not every person may experience sacred silence but whoever does can end it whenever s/he wants to. Getting drawn into the sacred moment has to be understood according to radical constructivism as an individual self-determined action. It certainly helps if the surrounding soundscape corresponds with the inner need for silence. Silence is part of the service. In contrast to the silence persons can end themselves, silence *staged* by the pastor or organist might be experienced as too short or long.

> When I'm at the piano and its right before the prayer I will look out and see what people are doing that 15 seconds before [...] pastoral prayer [starts]. [...] I'm wondering if maybe we could be longer to get into it. I love silence (Heenan 2011).

In order to not disturb believers from their silent moments, music is often designed to be non-disturbing (see above). But as we will see in a moment, even bold sounds might not destroy *sacred silence* but rather strengthen it.

Being silent is accepted behavior in a sanctuary and a good aid for knowing God (Barr 2012). It might be understood as "the most appropriate response [to God], because it is only in silence that any possible meaning can be found" (Zembylas and Michaelides 2004). From a phenomenological perspective the answer (or response) belongs to the *own*, while the question is always *alien*. Thus, it has to be discussed where a possible communication with God takes place and where sacred silence has to be located.

But do we need silence to experience sacredness? "They say listen to the little voice, the tiny voice, the small voice within your heart. When do you do that when everything is cacophonous? I think it's important, very important" (Heenan 2011). So, it seems that we do. As there is no acoustically measurable holy response to the *own* answer, being silent oneself appears to be a *normal* (re)action. The discipline of the ritual backs the need to be silent. There are rather strict traditions related to how certain parts of a service have to be performed. You can attend a Catholic mass in whatever country, held in whatever language: The process, including the cadence and rhythm of prayers like *creed* and the *Lord's Prayer*, is always perceptible and comprehensible. "There's something about that discipline of being silent that helps us open…shove other things out of our realm of what's going on and making room for that spiritual experience, that spiritual listening" (Edge 2011). There can be prayers hearable as part of the soundscape, but the ritualized performance does not disturb the silence of the believer. It supports the *inner silence*, a silence "in your thoughts. You can have a sense of silence. Out of silence […] in our mind the heart can commune with God" (Barr 2012). The silence "in your thoughts" must not be understood as a total absence of all thinking. It is more like letting go of all ordinary, everyday thoughts and focusing on the conversation with God. In the experience of sacredness, silence "becomes powerful. […] There's silence but there's not silence going on in one's head because one is contemplating or ought to. I think that's a very powerful worship" (McDonell 2011). As there are different kinds of silence, from measurable acoustics to the inner silence of a believer, the question of what *sacred silence* is has to be investigated further.

A first clue arises out of the empirical fieldwork conducted in Austin's downtown churches. Silent places in the urban environment are usually defined like every place else: via decibels. The discussion about *tranquil space* shows (see above) that green spaces are sought out by the urban dweller in need of quiet time. Buildings open to the public like libraries, museums, and churches are also recognized as locations where quiet can be found. So, how silent is it inside a sanctuary? In all churches where interviews where conducted, a decibel measurement device was installed to sample the dB(A) of the places. Generally, results lie between 30 dB(A) and 40 dB(A), a rather quiet environment compared to the outside soundscape in downtown Austin. The lowest value was recorded in Central Presbyterian Church, with 25.7 dB(A), while temporary heights of up to 41.5 dB(A) arose, caused by creaking in the wooden rafters. All churches surveyed have stone walls, most of them plastered, and most of them have carpet on the floor. The acoustics of Central Christian Church, the only church using stone flooring and having both rather big windows and two indicated side aisles, were affected strongly by traffic passing by the sanctuary. The general 38.5 dB(A) rose to over 50 dB(A) because of cars and trucks. To assume that believers in Central Christian Church might not be able to experience sacred silence in the same way those in Central Presbyterian Church would – because it is louder there – is absurd. To measure silence by decibels does not make sense in a sanctuary, where sound – including its absence – is highly individually felt, interpreted, and experienced. Putting up a decibel-measurement device in a sanctuary not only shows how ineffective this method is at describing the soundscape of a house of God, but any place. Silence, and especially one that is defined as being sacred, has to be of another quality than the mere absence of sound.

So, what is *sacred silence*? "I think that the silence is different. Even if you go into government buildings that were built with the same material, unless they were mimicking a church they didn't have that" (Stevens 2011). Silence in a place like a sanctuary is more than just low decibels. It is experienced as being purposeful, carrying certain meaning. "In the church I think you have more of a planned [silence]" (Barr 2012). The difference between the silence in places like the State Capitol, a library, and a museum and that in a sanctuary could be described phenomenologically as a difference of *order*, a difference of *own* and *alien*. While the silence in secular buildings is an understandable, explainable, and therefore ordinary one, sacred silence resists every understanding. Sacred silence belongs to the *extraordinary* of the alien (see Waldenfels 2011: 4). However, as "purposeful silence" (Barr 2012) it appears in the very place of the sanctuary, where it helps "to […] maintaining [religious] norms" (Bruneau 1973: 43) that are obviously hard to define. Unlike the *Orient*, whose alienness leads to stronger self-identification of the *own* (see Rose 1995, Said 2003), one characteristic of sacred silence is to surprise and disturb us (Stevens 2011), which makes it so hard to describe. It is like the known and *familiar* that is disturbed by an alien experience that we cannot explain nor really grasp. While everything *own* is located *here*, the *alien* is *there*. The interesting thing about a confrontation with the alien is that – at least phenomenologically – it does not have to exist *somewhere else* but can occur right within the home. The alien disturbs (Tuan 1986, Waldenfels 1997: 42) exactly because it appears not only far away but in familiar places, places we feel safe and secure in: like church.

> [T]he sounds you hear don't particularly have to be unique to the church. Maybe the combination of smells, visuals, and all that together make people feel like they're in a holy place (Stevens 2011).

The combination of sight and sound, or rather all senses for experiencing sacredness, is surely an important element for religious places. The senses serve as "powerful agents of breaching the thin veil between the natural and supernatural worlds" (Nelson 2007: 236). Thus, whenever silence in a sanctuary is associated with the Holy and used as a vehicle to communicate with God, find inner peace, or something comparable, we can speak of *sacred silence*. What makes people experience the holiness of the sanctuary is founded on individual perception based on every element of the own stream of consciousness (Waldenfels 2006: 118–119, see Husserl 1983: 363). "I think maybe silence…again the very physical silence I think might be different. The metaphysical silence certainly must be" (Stevens 2011).

There is no question that pure or absolute silence does not exist (Metzer 2006: 334) – at least in places like a sanctuary. Sacred silence cannot be defined as an absence of sound but more like a mental state. It seems absurd to perform decibel measurements to find out about sacred silence, when "even the services with the sound of the services create an environment for silence" (Barr 2012). You have to have sound to experience silence. In music silence is used to make possible the experience of sound, and vice versa (see Figgis 2003: 5). Waldenfels follows the dichotomy of *own* and *alien* implicitly in discussing silence. It is not just a phenomenon of not hearing anything, but of not hearing something. Sound is there latently when we experience silence. Individual perception makes us experience silence from our own perspective. The fact that something is not heard (by the listener) only speaks to the listener's cognitive abilities for experiencing sound. Silence is not simply toneless. It is there like text between the lines (Waldenfels 2010: 178–9). I argue that silence belongs to sound. It cannot be heard but defines itself by this very quality. Silence cannot be heard but is experienced within a soundscape that defines its *scape* via sound. Sacred silence, as a presence that "cannot be experienced with […] the normal sounds of life" (Barr 2012), is the *alien* of normal sounds. It is inherent to sound and thus part of the soundscape wherein it is inaccessible originally (see Husserl 1960). The *alien* of sound exists parallel to an *own* and can never be deducted from it (Waldenfels 1999: 184). Culture as a whole is a good example of this dichotomy. As discussed in 1.1, both own and alien culture develop simultaneously. The foreign language becomes alien the moment the mother tongue is learned and internalized as something familiar and belonging to the own. Thinking about sound and silence in a similar way, the latter is always part of sound while – per definitionem – it is not sound.

I would like to propose looking at silence as the alien of sound.

> A phenomenon like the alien, which shows itself only by eluding us, could be characterized as a hyperphenomenon. Accordingly, Husserl characterizes the alien as 'a verifiable accessibility of what is inaccessible originally' (Waldenfels 2011: 35).

This *inaccessibility* (see Waldenfels 1997: 25, 1998: 42) is exactly how silence in places like the sanctuary is perceived. "[S]ilence is […] it's not an absence. Silence is not an absence of, it's maybe an absence of noise, but it's not a negative thing. It can be a very positive thing" (Barr 2012). The absence of sound is something that is an inherent part of sound. "You can't hardly have it without it" (McDonell 2011). Waldenfels' understanding that "there is no sufficient justification or explanation for the experience of the alien […]" (Waldenfels 2011: 4) is the phenomenological description for the evaluation of sacred silence as something that is "a presence, but [...] cannot be experienced with kind of just the normal sounds of life" (Barr 2012).

Outside the sanctuary sacred silence can also be experienced. It is not necessary to be inside the house of God to sense an extraordinary – therefore alien – quality of silence. "Sacred silence is in lots of places, for me. Well, like nature, mornings on my back porch and places like that, take on sacred space for me, and sacred silence" (McDonell 2011). Mostly, a natural environment with less sound is described as a place where sacred silence might be perceived. The sacredness of silence is the alien aspect of a silent moment when God can be sensed or imagined. It is a silent exchange in between the known and unknown.

> I've experienced sacred silence when I've been walking down a cat trail in the woods or sacred silence when sitting on a hillside in a camp setting praying. I think sacred silence is any time when in the midst of that silent prayer we are cognizantly aware of God's presence with us and our desire to be in love and in relationship with God and to receive God's love and grace in our life. I would think that's sacred silence (Edge 2011).

The same way silence disappears during services whenever we consciously become aware of it and realize how silent it is and immediately start hearing various sounds surrounding us, sacred silence is nothing that rises out of the *own*. Experiencing sacred silence is something that "falls on you" (Stevens 2011). It is the alien question to the own answer (Waldenfels 1998: 95–6) that "seeks us out in our own home (German *heimsuchen*) [...] by surpassing our expectations and eluding our grasp" (Waldenfels 2011: 3) (see 1.1).

Since we are not able to describe sacred silence it is no wonder this silence is considered *alien*.

> The radical character of the alien does not mean that the alien is something entirely different from the own and the familiar; however, it does mean that it can neither be deduced from the own nor subsumed under the general (Waldenfels 2011: 35).

Experiencing silence is not something that has nothing to do with the listener. In a phenomenological sense it is all about the listener. S/he bears the center of everything own and is "the spatial Ego-center as the relational center of all spatial orientations and all possible presentations" (Husserl 2010: 109). As discussed in 1.1 the ego-center doesn't describe a static position, though it is potentially physical. The only positioning possible is a relational one, as this ego-centered core always moves with the *own* and stays in its center, in constant distance from everything *alien* (see Husserl 2010: 281). The facets that constitute the own in its individuality are constantly changing and adapt depending on apperception (Husserl 1995: 102). We already talked about the fact that:

> [i]f the essence of perception does not lie in apperception, then all talk of perception in regard to external things, mountains, woods, houses etc., is misguided, and this, the normal sense of the word *perception*, surely illustrated in these cases above all others, must be abandoned. Outer perception is apperception [...] (Husserl 2001: 341).

Thus, *own* and *alien* are interlaced and can never be differentiated clearly from each other (Waldenfels 1997: 67). Without the *alien* there would be nothing to call *own* because of the missing relational counterpart. This is exactly what archpriest David Barr means when he says: "Sound and its absence both are a very important part of worship and helping us to know God" (2012). Co-pastor Greg McDonell agrees with this belief that sound and silence go hand in hand and cannot be separated (McDonell 2011, see above). The urgent phenomenological differentiation of *own* and *alien* includes not only a certain distance and a radical closeness of the two; it also bears a more dynamic quality: The own's wish to extend its familiar territory and to push the alien as far away as possible (Waldenfels 1994: 423).

> As Ego in the transcendental attitude I attempt first of all to delimit, within my horizon of transcendental experience, what is peculiarly my own [Mir-Eigenes]. First I say that it is non-alien [Nicht-Fremdes]. I begin by freeing that horizon abstractively from everything that is at all alien (Husserl 1995: 97).

But Tuan's believe that "[a] principal aim of culture is to extend the realm of the familiar at the expense of nature and the strange" (Tuan 1986: 16) seems to be questionable in the context of sacred silence and the Holy in general. Sacred silence is nothing we would define as *own*. As a way to experience God sacred silence stands on the other side of the border between *own* and *alien*. It is something *extra-ordinary* that we explicitly seek and use as a "doorway" (see above) to experience God – who is the very example of the extra-ordinary Holy. But while sacred silence is characterized as being *alien*, it can also be experienced. It carries a core of the own (Husserl 1973: 432). In sacred silence we are affected by the *alien*, while every form of experience belongs to the *own* (Husserl 1995: 102). Perception – as stated before – is an active process, founded on the individual stream of consciousness. Confrontation between *own* and *alien* takes place in between known and unknown, familiar and unfamiliar, in between here and there, ordinary and extraordinary (Waldenfels 1997: 73, 1998: 45–6). Sacred silence describes the individual experience of sound, where the threshold between sound and silence is less important than the dichotomy of *own* and *alien*. Sacred silence cannot be measured because it constantly withdraws from being absorbed by the ordinary.

Chapter 5
Excursus: The Medial Creation of the City in Audio Drama

5.1 Sound in Media

Environmental impressions are not the only ones that have an impact on how we perceive the world. As the example of audio-guided tours in Chapter 4, section 4.4 has shown, media also plays a part in co-shaping what we experience. In the case of the audio tour, aside from the storyteller there is always the urban environment through which a listener walks while listening. As described above, this environmental influence has to be considered when designing audio tours. But there is also media that doesn't rely on the surroundings of its consumer. Film, for example, creates a unique sense of place that has to be compatible with the ideas and believes of the recipient. The surroundings of the location where a movie is being watched are generally not considered (an exception might be the on-location viewing described by Zonn (2007)). Generating a sense of place in media is nothing new to geography (Adams 2009, Aitken and Zonn 1994, Döring and Thielmann 2009, Rose 2012, Schlottmann and Miggelbrink 2009).

> Recognizing geography as a visual discipline – in theory and in practice – has become, in the meantime, the embodiment of an entire wave of research and publications (Wissmann and Zimmermann 2010: 372).

Most authors pick visual sensations as "focus of media geographies" (Pesses 2009: 147) (see Escher 2006, Lukinbeal and Zimmermann 2006, 2008). Sound is generally left out, though Tuan already stated in 1994:

> What radio and television stations report or what newspapers print can make a large difference in how people see, behave, and act in a city (1994: 148).

This excursus assumes that (mass) media strongly impacts the creation of sense of place (Zimmermann 2007), and "frequently makes use of more than visuality" (Wissmann and Zimmermann 2010: 371). Audio drama and audiobook have "become increasingly popular in the past few years due to the ubiquitous availability of digital media" (Wissmann and Zimmermann 2010: 371); the following investigation uses audio drama to show how audio media in general has a cognitive impact on the listener without directly using the environment. The basis premise is that there always have to be connections between the sounds in media and the familiar impressions of the listener's life (Pesses 2011: 42). Otherwise, it would not be possible to deconstruct audio media, i.e., is to understand what we are listening to. With the generalization and simplification of sounds playing an important role in audio creations of the world in audio drama, the question is how sense of place is produced solely through audio. To answer this question the following section will take a closer look at two popular cultural audio drama series, *The Three Investigators* and *Gabriel Burns*, that take similar and yet different approaches to generating sense of place.

5.2 Audio Drama – *The Three Investigators* vs. *Gabriel Burns*

Audio drama is not a new genre. Originally recorded for radio – and accordingly called 'radio drama' – today's productions are sold on sound carriers like CDs and downloaded digital via the Internet. They "incorporate staged elements and are thus acoustic narrations that go beyond spoken texts or words read aloud" (Wissmann and Zimmermann 2010: 374). Germany is the most important market for audio plays – this is what single audio drama series are called –, with more than 11.6 million copies sold in 2009 alone (Sony Music Entertainment Germany GmbH). The most successful production in Germany is also the world's most popular audio play, *The Three Investigators* (*TTI*), with over 44 million copies sold (Neumann).

> To put this in context, the more than 40 million copies sold matches the combined sales of the five best selling music artists worldwide for Germany (The Beatles, Elvis Presley, Michael Jackson, ABBA and Madonna) (Wissmann and Zimmermann 2010: 373).

Other productions have been around for similar lengths of time and/or also sell in significant numbers. EUROPA, part of Sony Music, is the largest producer of audio plays, focusing mainly on children. Aside from TTI, which started out as a children's audio play and today mainly attracts adults between 30 and 40 years of age, action, adventure, and, most of all, detective novels, are most popular – like TKKG, which sells about one million copies annually to its target group of eight to 10 year olds (Sony Music Entertainment Germany GmbH).

The detective novel may be the most successful type of audio drama because it – like its written cousin – "enables the representation of a special view of society and contemporary perspectives [...] by using generalized and simplified characters" (Wissmann and Zimmermann 2010: 374) (for a closer analysis of the works of Agatha Christie, Arthur Conon Doyle, and others see Bloch 1965, Eco 1998, Heißenbüttel 1966, Revzin 1998, Sklovskij 1998, Todorov 1998). Siegfried Kracauer (1979: 22) believes that in the detective novel recipients can get a clearer look at society than by observing the actual environment. The "novel provides society a clearer picture of itself than it would be perceived through their every-day experiences" (Wissmann and Zimmermann 2010: 374). *TTI* tells the story of three male teenage detectives who live in fictitious Rocky Beach, a town on California's pacific coast between Hollywood and Santa Barbara. From their headquarters located in a salvage yard owned by of one of the boys' uncles, the boys solve mysteries; as of September 2013, there had been 163 episodes in the series. Since the series' birth in 1964 (Akstinat 2008: 9), when the first books were published in the USA, the popular-cultural impact of *TTI* has become obvious, not only in sales figures and numbers of episodes but in the whole media cosmos that spins around the audio play. There are board games and digital games, numerous Web sites, and even live performances by amateur drama groups. In a large online survey, Wissmann and Zimmermann (2010) learned about fans getting married based on *TTI*, taking trips to the imaginary Rocky Beach, and naming their children after the lead characters Justus, Peter, and Bob. Various scientific final theses have been published focusing on audio plays, especially *TTI* (Haaf 2003, Köchy 2005, Rodermond 2005, Schmidt 2002).

According to Kracauer's idea of using simplification and generalization to better explain and recognize social structures (1979), the audio production of *TTI* makes use of only few sound objects (see Schafer 1994: 274). All lead characters are still voiced by the original actors, and many sound objects have been re-used since then. A totally different approach comes from the producers of *Gabriel Burns*, a mystery/thriller audio play located in the Canadian city of Vancouver. Since starting the show in 2003, producer Volker Sassenberg has tried to incorporate always-new sounds. Internal guidelines speak about generating unique sound objects for everything hearable in the audio drama series (Bruß). Unlike in *TTI*, where there have existed only two different sound samples of a barking dog since 1979, in *Gabriel Burns*, every dog, door, or car has to sound different. As will be discussed in sub-section 5.2.3, this does not automatically lead to a richer soundscape for a single scene, but it surely enhances the complexity and originality of the production. Official sales figures do not exist for *Gabriel Burns*, but audio drama charts show that the series is not unsuccessful. With nine places in the German audio drama top 10 chart for 2010, *TTI*'s absolute dominance was only broken by *Gabriel Burns*' episode 34 (I know what Fear is), which ranked 8th (Göldner). Like *TTI*, *Gabriel Burns* appeals mainly to adults. The lead characters, Stephen Burns, Larry Newman, and Joyce Kramer and their boss, Mr. Bakerman, fight alien threats and are confronted with magic, mystery, and horror. In a setting that constantly becomes darker and less hopeful, the ultimate goal is to prevent earth from falling into the hands of the mysterious Pale ('die Fahlen'). When mysterious or magical events occur, it is interesting to hear how normally non-existent sounds are generated to evoke an intended reaction in the listener.

To find out about how both audio drama series translate the world into audio, I undertook a soundscape deconstruction. The first 30 episodes of *Gabriel Burns* (that is the first season of the series) were contrasted with 40 episodes of *TTI* (where, starting with episode two, each fourth episode was picked to stand for the whole production range). Based on the approach of the World Soundscape Project (see Schafer 1994), audio was categorized into *keynote*, *signal*, and *soundmark*. As 4.4 has mentioned, Wissmann and Zimmermann (2010) added another three categories to fully cover all sounds occurring in audio drama: *storyteller*, *soundframe*, and *dialogue*. As in audio-guided tours, spoken content dominates the soundscape of audio drama. *Storyteller*, as "all-knowing narrator to present the story", and *dialogue* between characters "provide references to the environment of the persons in the story" (Wissmann and Zimmermann 2010: 376). Thus, it made sense to start listening to the ability language has to evoke the world. Section 5.2.1 will start with a text-based analysis of audio plays' scripts. As the main focus of this book is the geographies of urban sound, evocation of the city will be of primary concern. Associations with both Rocky Beach and Vancouver are analyzed, which will tell more about how listeners are stimulated. Aside from associative classification, another way to evoke sense of place in audio drama is to deconstruct places of action (2.5.2) (see "chronotopoi" of Bachtin 2008, Wigbers 2006). Depending on the focus of the study, those places can be classified in relation to the characteristics of the detective novel genre (Wissmann and Zimmermann 2010: 377) or, in the case of the inquiry of this book, regarding their functions within the urban environment. Following the functional classification of the city, Sub-section 2.5.2 will tell more about how the city is represented functionally, and if there might be more or less typical places of action to evoke an intended image of the city. Section 2.5.3 leaves *storyteller* and *dialogue* behind and concentrates on the last four soundscape elements, *soundframe*, *keynote*, *signal*, and *soundmark*. Classification through sound tells about generalizations and stereotypes in audio drama, which are present not only in *TTI* but *Gabriel Burns*, too. How many sound objects occur simultaneously? What are the dominant soundscape elements aside from *storyteller* and *dialogue*? Can community sounds exist within an audio production that is *only* imaginated and where no community exists to define a sound as a *soundmark*? And what about the total absence of sound? It will be interesting to hear if the statement discussed in section 2.4 is still true:

> If you want silence, you have to approximate silence with what's called *room tone*. It's like quiet white noise. But you can't have zero (Figgis 2003: 1).

Sub-section 2.5.3 will show that there are various ways to use sound (and its absence) to evoke the city without being dependent on spoken language.

Regardless of the way soundscape in audio media is deconstructed, it always has one single purpose: to stimulate the listener who integrates those stimuli in the individual stream of consciousness where they influence the way we perceive the world.

5.2.1 Associative Classification

Words are a powerful way of stimulating the listener. The popularity of audiobooks shows that there is a mass market for audio media that solely uses language to narrate stories and evoke places. While audio drama makes use of more than just words, both *storyteller* and *dialogue* hold dominant positions within the six soundscape categories. As the plot evolves and the characters move from one place of action to the next, information is given about the environment where the story takes place. A text-based analysis of the audio plays' scripts can reveal how certain places are described and connoted. There is no difference and thus no differentiation between *storyteller* and *dialogue*, as they are both responsible for evoking place narratively. "The storyteller [...] lends a hand in creating the life world by verbally describing the places of action with the included elements" (Wissmann and Zimmermann 2010: 376), while "*dialogue* provides references to the environment of the persons in the story" as well (Wissmann and Zimmermann 2010: 376).

To find out about how the major city of each audio play is described, all sections are extracted from the scripts that contain either the word *Vancouver* – for *Gabriel Burns* – or *Rocky Beach* – for *TTI*. In addition all passages containing districts, suburbs, and other parts of Vancouver and Rocky Beach are included in the analysis. Of the 30 episodes of *Gabriel Burns*, seven do not contain references to Vancouver, while in six of the 40 *TTI* episodes Rocky Beach is not referenced. Within the selected passages there are three different types of information given about the city: places of action, references to the environment, and atmospheric language associated with the city. Words describing places of action, such as *hospital*, *Chinatown*, and *living room* are recorded for a closer analysis in 5.2.2. References to the environment give a first impression of the city. Rocky Beach is associated with the Pacific coast, a nearby national park, and the Santa Monica Mountains. With the overall setting in California, these three references are sufficient to locate Rocky Beach and simultaneously set the scene for 40 *TTI* episodes. That ostensibly meager result matches Kracauer's understanding of the detective novel world, where little information is used in order to create a sharp image of the world (see Kracauer 1979). As described above, the guidelines for *Gabriel Burns* are different. The targeted richness of sounds can also be heard in how Vancouver is described. There are 14 different environmental references, occurring in 27 single passages in the 30 *Gabriel Burns* scripts. The most frequent mention is *Vancouver Island*, followed by the *coastal mountains*, and the nearby *forrest*. In addition, the city's land is described as *ancient*, there are references to *coast* and *ocean*, *bay* and *branch of a river*, and a *tarn* and a *valley*. The description is way more complex than that of Rocky Beach, which is also true for the atmospheric language.

> Storyteller: Black mist had made the world sodden and robbed it of all its shapes. A pale moon sat over the city in between bloodless stars. The streetlights had gone out. A car stood diagonally on the street, the co-driver's door open. Nowhere there was light. Nowhere a sign of life (*Gabriel Burns*, episode 30, "White", time code 11:50–12:29, translated from German).

The setting in the example is rather mysterious and dark. Vancouver is described with words like *bloodless* and *pale*. The description creates a world suitable for the overall story of *Gabriel Burns*. The mystery/thriller genre is clearly represented in Figure 5.1, where both Vancouver and Rocky Beach are displayed via accompanying atmospheric language. The bigger words appear in the figure, and how often they occur in each audio play. There is a rich description of Vancouver and the actions that take place in it. Associations like *dead*, *night*, *pale*, *darkness*, and *fear* absolutely reflect the audio play's genre. While there are 174 single words evoking a certain atmosphere for the city in 30 episodes, *TTI*'s Rocky Beach and related plots are portrayed less colorfully. About half the associations (86) spread over 40 episodes of the detective novel audio play also strikingly display what the genre is all about. *Robbery*, *valuables*, *night*, and *secret* are the most named words, supported by typical genre keywords like *offender*, *hunt*, and *interrogation* (see Figure 5.1). It almost seems that those latter three words tell the whole story of a classic detective novel: Offender – hunt – interrogation.

The fact that words associated with a place like a city in audio drama strongly refer to the production's genre suggest that language is used more to communicate a specific atmosphere than describe the urban environment. It might be an interesting question to ask: whether this is always true in media. If the genre determines how place is represented in media, what would this mean for studies about representation of places in media? In the case of audio drama and the creation of the city through sound, language seems to be insufficient for fully communicating place.

> Storyteller: In the old theater there was nothing suspicious to the three detectives. Thus, they drove back to Rocky Beach without getting a step further. They decided that Justus and Peter should stay in the house of Miss Agawam for the night. They hoped to suss out the gnomes if they should appear (*TTI*, episode 22, "The Mystery of the Vanishing Treasure", track 4, time code 00:17–00:41, translated from German).

The example shows that language is also used to summarize the story "including a recapping or preview of events" (Wissmann and Zimmermann 2010: 376). In addition to communicating leaps in time and accentuating narration, the dominance of the *storyteller* is used in various ways to keep the story going. *Dialogue* most of the time consists of character-to-character conversations regarding the plot, solving problems, and discussing actions. Sense of place of a city is co-evoked by those soundscape elements but only in addition to other sounds of the production. Before we focus on the latter, the next section will look at the places of action named in *storyteller* and *dialogue* to understand how single locations can add up to create the city in another way.

5.2.2 Functional Classification

After a most common *soundframe* in the beginning of an episode – comparable to the opening of a television show – the story of an audio play begins. Either

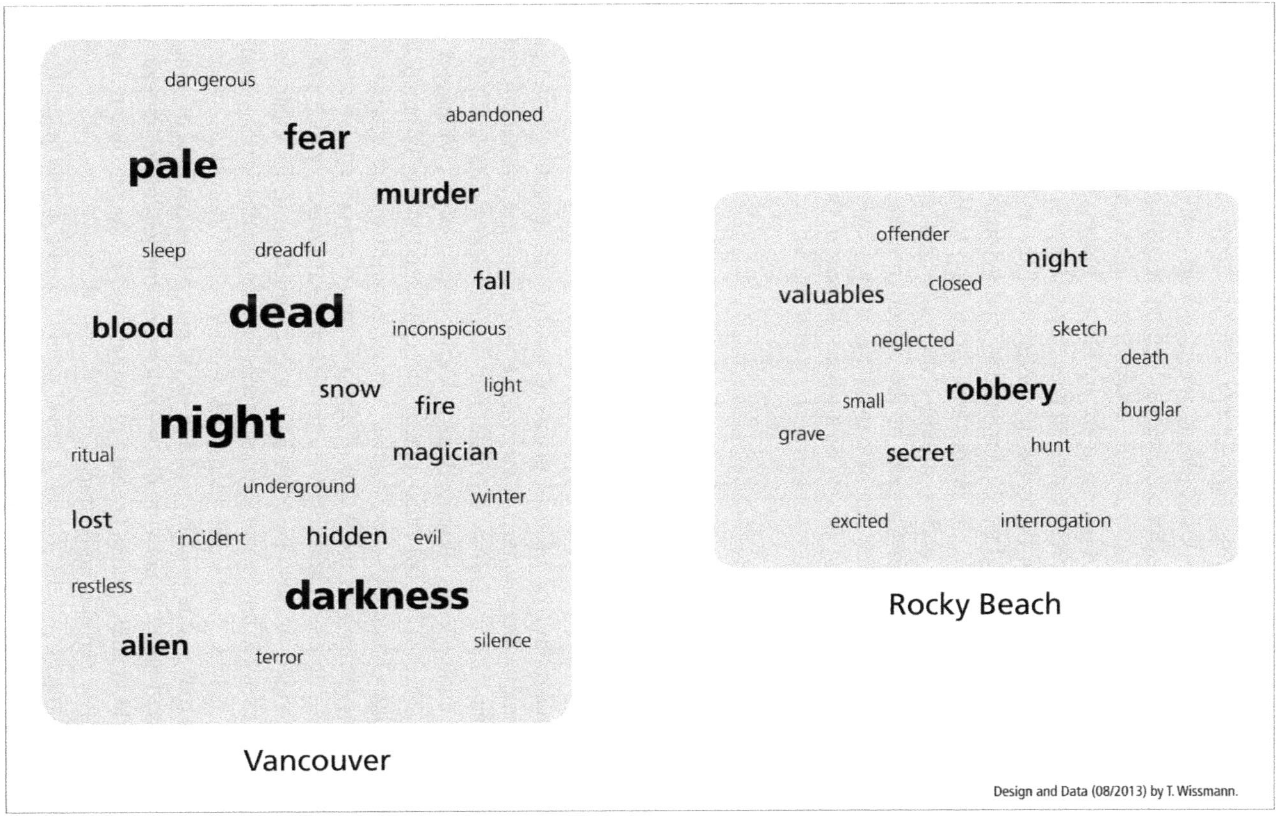

Figure 5.1 *Gabriel Burns* **and** *TTI* **– Associations Clouds**

the storyteller sets the scene for the upcoming events or the characters directly act within their environment. Location in audio drama is of high importance. It makes up almost all of the time of the recording. *Soundframe* and possibly *storyteller* are the only two soundscape elements that withdraw from the constantly evolving plot. In the case of *TTI* only about 10 percent of the time *soundframe* interrupts the series of places of action – generally with the purpose of better tying them together (for a closer description of *soundframe* see 5.2.3.1). The question about where the story in audio drama plays takes place can be answered in several ways. Of course, there are always places, like an *apartment*, *street*, or *car*, where the characters are, but how those *places of action* are defined and categorized depends on the epistemic interest of the respective study. To deconstruct the *TTI* audio play, Wissmann and Zimmermann (2010) focus on the detective-novel aspect of the production. According to Bachtin (2008) and Kracauer (1979) places of action structure the story and determine both the space and time of action. In the case of the detective novel there are several (stereo)typical places of action that help to tell the story, like *hotel lobby*, with its ignorant visitors representing society as a whole (Kracauer 1979: 48). Places of action help the detective locate himself in the world and act according to specific rules and norms.

> Taking apart the description of the places of action shows just how important those locations are for the underlying setting in addition to the detective story. While the presentation of the setting often fades into the background, there are times when it cannot be ignored (Wissmann and Zimmermann 2010: 377).

The authors make out five different places of location for detective novel audio drama: *Independent location*, *headquarters*, *scene of discovery*, *consecutive place of discovery*, and *place of solution*. Aside from *independent location*, where "the world of the audioplay is described particularly intensively" (Wissmann and Zimmermann 2010: 377), all other places are connected to the criminal case. Locations – like characters – are labeled and stereotyped by the listener. They develop an impact on how the world of audio drama is perceived. Cross-media experiences help to create and foster those stereotypical images, which virtually extend the provided soundscape of audio drama production. If, for example, the three detectives walk into a salvage yard, the listener automatically constructs this place of action. S/he uses memories from his/her individual stream of consciousness that enable her/him to fill in the blanks where audio is missing. As will be discussed in 5.2.3, individual associations are tremendously important to create the world in audio drama because of limitations like the maximum number of simultaneously playable audio tracks.

The deconstruction of the detective novel genre in audio drama helps listeners understand the importance of places of action. Apart from bringing to life a crime story, places of action always evoke the associations of the listener. If the story takes place in a car, for example, it makes a huge difference if the *storyteller* describes it as a *cab*, *pickup truck*, or *race car*. Without the ability to hear each motor's engine, the listener automatically uses individual imagination and personal experience to evoke this place of action. Individual perception of sound is not only true for environmental sounds – as described in 4.2.3, using the example of air traffic noise – but also sound in media, even if it cannot be heard while listening to the story.

As the classification of settings described by Wissmann and Zimmermann focus on the story's content, and is limited to the detective novel genre, places of action have to be divided up into other categories. With the book's overall interest in the urban environment and the author's aim to find out how the city is created in audio drama, places of action are classified according to the common (German) zoning plan. In this plan there are several zones structuring the city functionally. The type of usage of a building depends on the zone where it is located. A commercial agency, for example, would not be allowed to be located in a pure residential zone. To apply this classification to an audio plays' places of action each of the combined 70 episodes are deconstructed into a huge number of single places. For the 40 *TTI* episodes 477 places of action are counted, similar to *Gabriel Burns* (481). As the sample of the latter (30 episodes) is significantly smaller, characters visit about four places of location more per episode (16.03) than in *TTI* (11.93). After all places are listed in phase one of the analysis, in phase two they are filtered according to their location. Only those located in Vancouver and Rocky Beach are used. According to the zoning plan, where usage of each zone is described in detail, each place of action is assigned to one of the following nine zones: *commercial*, *green area*, *special*, *traffic*, *community facilities*, *water*, *residential*, *excavation and mining*, and *common use*. Places of action of the last zone, *common use*, were combined into the *community facilities* class, while *excavation and mining* was integrated into the *special* class. Thus, relatively similar usage like *tunnel* (originally belonging to *excavation* and *mining*) and *vault* (belonging to *special*) were combined logically. As the table in Figure 5.2 shows, most important in *TTI* are places belonging to *commercial* (35.03 percent) and *residential* areas (30.51 percent).

A look at the accompanying sound cloud reveals that the three places of action occurring the most belong to these categories (as in Figure 5.1, places of action occurring more often are displayed by larger font and more dominant style). The three detectives spend most of their time in their headquarters (*commercial*), an old trailer in Justus's uncle's salvage yard.

> The article about the three detectives came out only the next morning. Bob read it to them as they gathered in their headquarters. The article stated their name and explained them having their central office at the salvage yard of Jonas Company in Rocky Beach (*TTI*, episode 6, "The Mystery of the Talking Skull", time code 04:20–04:36, translated from German).

The setting in the middle of a salvage yard – itself the second most important commercial place of action in the *TTI* – communicates a kind of chaotic, hidden,

Vancouver

Rocky Beach

Number	Function	Three Investigators	Gabriel Burns
1	Commercial	35.03%	13.61%
2	Community Facilities	13.56%	12.57%
3	Green Area	4.52%	5.24%
4	Residential	30.51%	29.32%
5	Special	1.13%	12.57%
6	Traffic	14.12%	24.61%
7	Water	1.13%	2.09%

Design and Data (08/2013) by T. Wissmann.

Figure 5.2 ***Gabriel Burns*** **and** ***TTI*** **– Locations Clouds**

and special meaning for the detectives' office. In a way it matches the stereotypical headquarters of the classic Chandler-esque detective novel in some upper floor or back part of a shabby building, where an office is filled up with a mess of papers and empty glasses of whiskey sitting on numerous pieces of old furniture. As there is no in-depth description of the trailer's interior, the setting itself is supposed to evoke place. The second most important place of action in *TTI* is the *entrance area*, including the *foyer* and – not unexpected for a detective novel – the *lobby*. It combines the entrances of not only hotels and restaurants but also residential buildings. The latter is even more dominant than commercial buildings, so that it is classified as belonging to *residential* zone. The same goes for *house*, the third most important location in *TTI*. Many clients set up meetings with the detectives at home, while the detectives also solve their cases on private property.

> Today, the city owns the house. It seems that until now nobody got around to taking a look at it. Because the death of Billy Ford is no secret, it could be that an ordinary robber wanted to rip off some valuables as long as the house stands still empty (*TTI*, episode 74, track 1, "The Burning Sword", time code 06:10–06:24, translated from German).

As the cloud in Figure 5.2 shows, Rocky Beach is also dominated by *traffic* places of action like *street*, and *car*, while the rest of the functional classes appear less often. Two of them are almost totally missing: *Water* and *special.* At least places of action belonging to *water* could be assumed to play a bigger role in a city that is located on the Pacific coast – and even carries *beach* in its name – than *harbor* and *bay*, both of which are mentioned only once in all 40 episodes. But as with the associated classification in 5.2.1, Rocky Beach is described not by numerous extraordinary locations. Places of action like *entrance area*, *house*, *car*, and *street* within the top five locations do not tell a specific story. We learn that the city has a *police station*, a *theater*, a *museum*, *circus*, and a *graveyard.* But there are no special locations that could be seen as being unique features. In the tradition of the detective novel, Rocky Beach and its locations appear to be a small version of the world in which the three detectives act as moral conscience.

Vancouver has more to offer. Places of action can be found in all seven functional categories. Thirty-five places get mentioned at least twice in all 30 *Gabriel Burns* episodes – compared to only 20 places in 40 *TTI* episodes. While *residential* locations occur almost as often as in Rocky Beach (29.32 percent, see table in Figure 5.2), *traffic* is the second most important category (24.61 percent), containing the two most occurring places of action, *car* and *street.* A major part of the story takes places in between fixed locations. The characters' high mobility serves the overall theme of the mystery/thriller plot, where nothing seems to be reliable, persons and even locations change, and the world struggles between good and evil.

> Snowflakes fell soft and silent and put a dress of winter on the city's streets. The dawning morning announced a new day. The ambassadors of the Two Horizons were gone. The traffic light at the end of the street constantly changed its lights (*Gabriel Burns*, episode 30, "White", time code 46:26–47:05, translated from German).

The higher number of places of action in general means that the listener gets a more detailed idea of the city. There are not only *rooms* where the story takes place, but a wide variety of them: *Washroom*, *changing room*, *hotel room*, *bedroom*, *living room*, *bathroom*, and *hospital room.* There are also *special* locations like *cellar*, *prison cell*, and *vault* that add to the mysterious atmosphere of the story.

In comparison with Rocky Beach, Vancouver certainly is a far more complex city with more specific descriptions of all its various places of action. It appears to be less easy to deconstruct by the listener, particularly because of the way *keynote*, *signal*, and *soundmarks* are used to evoke such places (see 5.2.3). The complexity of the mystery/thriller genre in contrast to the detective novel can be experienced by just taking a look at the different kinds of places of action. While Rocky Beach speaks of *headquarters*, *lobby*, *house*, *car*, *salvage yard*, where *robberies* of *valuables secretly* occur at *night*, the city of Vancouver hosts *evil*, *dreadful*, and *hidden murders*, *rituals*, and *terror*, occurring not only in *houses*, *apartments*, and various *rooms*, but also in *alleys*, *vaults*, and *prison cells*.

5.2.3 Classification Through Sound

How do places of action sound? In addition to *storyteller* and *dialogue* there are four more categories of audio drama soundscape: *Soundframe*, *keynote*, *signal*, and *soundmark.* Based on the focus on the urban environment, only sounds from Vancouver and Rocky Beach were considered for analysis. From places of action (see Figure 5.2) 30 locations were singled out that occur in both audio plays. Based on the number of mentions, the most important places of action from each functional category were selected. *Commercial*, *green area*, *special*, and *water* all contribute one place (that is *office*, *graveyard*, *cellar*, and *harbor*). From the remaining three functional categories (*traffic*, *community facilities*, and *residential*) six places of action are selected (*car* and *street*, *hospital* and *police station*, *apartment*, and *house*). *Entrance area* is spared because of the high variety and differences in descriptions.

5.2.3.1 Around places of action – soundframe

Soundframe is different from *keynote*, *signal*, and *soundmark.* It is not part of the story like background music, which belongs to *keynote*. As described above (see 5.2.2) *soundframe* helps to structure both story and production with acoustic breaks. The main title of each audio play signifies the beginning of an episode. It is detached from the specific story but gets the listener into the mood of the audio play. *Gabriel Burns*' introduction, for example, is designed like the beginning of a fairytale. *Storyteller* accompanies *soundframe* to create an atmosphere of magic, mystery, and suspense. The main title generally doesn't change, while *TTI* producers had to replace the original theme because of legal conflicts. *Soundframe* in between scenes – most of the time synonymous with a change of places of actions – varies, but there are titles that reoccur in various episodes. Some *soundframes* are pretty popular. *TTI* episode 29 ("The Original Music"), for example, does not contain a new detective story but only the original score to the series. *Soundframe* can be rather non-disturbing and more like ambient background music. As such, it allows the listener to go over the story again and think about what s/he just heard. *Soundframe* can also be extremely attention-grabbing, loud, and even disturbing. There are passages where the story is interrupted abruptly and the music does not have any connection to either the previous or upcoming scene. Especially in *TTI*, some of those examples can be understood as an expression of the time in which a particular episode was produced. While it was common in the 1970s and earlier to use classical music and instruments of the orchestra to enrich the production (Beurmann 2011: 156) (*TTI*, episode 2, "The Secret of Phantom Lake", 1979), music genres changed to *jazz* (episode 14, "The Mystery of Monster Mountain", 1980), *rock* (episode 42, "The Case of the Weeping Coffin", 1987), *blues* (episode 50, "Thriller Diller", 1990), and even *techno* (episode 74, "The Burning Sword", 1997). Today, productions are more traditional again, citing *soundframe* from the beginning of the series (episode 142, "Deathly Ice", 2010), and are composed using classical instruments again (episode 152, "Skateboard Fever", 2012).

While the purpose of *soundframe* is to open and close an episode, and, as an interlude, signify changes of places of action, it can also be used to extend the plot. In *TTI* episode 126 ("Horror from the Moor") Justus is awoken by Peter calling him from a museum, where he's sitting through the night to catch some burglars (time code 39:21). While Justus is sleepy and doesn't really realize who is calling him, Peter is extremely excited and scared. A body preserved in marshland and usually displayed in a glass cabinet has vanished. Peter has barricaded himself in the control room, as he believes the dead body has left the cabinet itself and now walks the museum undead. He is frightened and won't leave his hiding place until Justus and Bob are there to back him. Justus, listening to what Peter says, is totally awake now and ready for action. He plans on picking up Bob to reach the museum as fast as possible. The scene ends with Justus energetically hanging up the phone and music setting in (time code 40:43). The *soundframe* begins with two high, slightly discordant notes, a triangle constantly playing, and the six-note motif played by low strings (probably viola). The motif is fairly driving, supporting the scene of Justus making a move. Some high notes of a xylophone add to the *soundframe*, and a single bell on a tambourine can be heard, played out of time to the beat. As the scenery changing from a boy's bedroom to a museum where a potential zombie is running around, the *soundframe* gets more surreal and abstruse, when a *Pulp Fiction*-esque electric guitar chord is stoke and the viola gets more forceful. Two single sound effects break the harmony not only figuratively (one slight touch of cymbals and a reverberating wah-wah organ). The motif changes slightly to end on a higher pitch as the beat of a kettledrum marks the change of places of action. The following spinet has nothing to do with Justus' bedroom but is a musical expression of the mystery going on in the museum at night (time code 40:57). The moment the spinet sets in and the place of action changes is also marked by a change of recorded tracks (from track 7 to 8).

While the use of *soundframe* to extend and reference the story is rarely used in *TTI*, in *Gabriel Burns* it is most common. The *soundframe* is also faded into places of action, where it becomes part of a scene's *keynote.* Sometimes, it is ducked and *dialogue* or *storyteller* can be heard before *soundframe* becomes dominant again. It is used at the beginning of track 4 of *Gabriel Burns* episode 7 ("Fangs of the Wind") (time code 12:30) to introduce the upcoming place of action. Bakerman is in Libya, discussing an expedition to *Kufra oasis group* in the south of the country with the vice-governor of *Benghazi*, *Palladino*, and Italian *Colonello Amato.* The men try to convince Bakerman to join the expedition, because their former archeologist has recently been murdered. When Bakerman finally decides to take the position as lead excavator, the *soundframe* starts to signify the beginning of the trip. The bells of an oriental *rigg* can be heard, followed by a *darabukka*, drumming the beat. A pipe contributes a melody, accompanied by a synthesizer. Shortly before the *storyteller* begins to talk about the travel preparations of the convoy, an *oud* sets in (time code 12:56). This string instrument plays another, attention-grabbing melody, which is ducked but not faded out. It becomes part of the *keynote* to support the atmosphere of the scene, which involves soldiers wearing torches, hurrying to prepare everything for departure. About half a minute into the scene (time code 13:28) the music fades out.

The above theses, that "you can't have zero" sound in media productions (Figgis 2003: 1), can be refuted with *soundframe* in audio drama, as absolute silence in the recording is also used to strengthen a story's atmosphere. In *Gabriel Burns* episode 8 ("Fog Lake") Stephen is diving in a lake near Desmond. People, even children, mysteriously disappear in the woods around the lake, and there are stories of evil living on the bottom of the water. Down below dead bodies float around Stephen, who all of a sudden gets ripped off his breathing hose. Panic sets in. Stephen is disoriented, caught in the middle of the corpses, and getting drawn deeper into the lake. Dramatic music sets in that increases until it is almost overmodulated. Then, to increase the effect of loudness and horror even more, absolute silence sets in. It lasts eight complete seconds (time code 56:31–56:39) – an extremely long time for an audio production. The total absence of sound gives the listener the chance to calm down and process the information that the main character may have just died.

Whether *soundframe* structures the production or supports the plot depends not only on the producer but the listener as well. *Soundframe*, as all of the soundscape elements, evokes an always-individual response to what is being heard, based on the listener's stream of consciousness. How *soundframe* can alter audio drama's atmosphere can be experienced when listening to the original and the most recent

TTI productions. Because of legal proceedings, producer *EUROPA* had to remove some of the *soundframe* from the first 40 or so *TTI* episodes. The music has been replaced, though a priority was not always placed on creating a similar soundtrack, as the example of *TTI* episode 18 ("The Secret of Skeleton Island") shows. While in the old version a rather light but dominant *soundframe* starts while Peter's father talks to his son and leads over to the next scene (old episode 18, time code 10:29), in the new version the *soundframe* only starts after the father has finished talking (new episode 18, time code 11:35). Both the two-seconds time difference – the remaining time code difference is caused by a new opening title and cutting – and the more mysterious and less disturbing *soundframe* radically alter the listeners' impression.

Soundframe is a rather vague soundscape element for evoking places of action. Especially when using popular cultural music genres neither effect nor future perception can be foreseen. Linking *soundframe* to the story and thus extending both story and atmosphere of places of action seems to be a good way to ensure high acceptability among listeners.

5.2.3.2 Location and attention – soundmark and signal

In contrast to *soundframe*, *keynote*, *signal*, and *soundmark* occur inside scenes and thus help to evoke sense of place. While *keynote* can be heard almost the entire time, *signal* only occurs occasionally. *Soundmark* is no more than an exception within the soundscape, and is missing entirely from all *Gabriel Burns* episodes (see Figure 5.3). The most important and best-known *TTI soundmark* is the *cawing of a myna bird* in the detectives's *headquarters*. While that location was not selected for comparative analysis – because there is no *headquarters* in *Gabriel Burns* – the bird nevertheless has to be mentioned. It is the most dominant sound in the whole *TTI* soundscape and signifies the series as a whole. In the audio play's first episode ("The Mystery of the Stuttering Parrot") *Blackbeard* appears as one of some parrots that have been trained to talk. Each parrot's line of text contains hints to solve a riddle. Blackbeard – actually a myna and not a parrot – is the only bird that knows the complete text. In episode 1 (time code 27:11) a woman walks into the house of Carlos and his father, two stereotypically poor Latinos, who had sold the birds for five dollars each. The woman is returning Blackbeard because she is not satisfied with the bird as a pet. As soon as the three detectives take a look at its cage, the myna starts to *caw* (time code 27:45) and, a few seconds later, begins to introduce itself: "I'm Blackbeard the pirate. I buried my treasure in the dark of the night, where the dead keep watch eternally" (time code 27:59). Here, the *cawing* is a *signal* for the bird. After Justus, Peter, and Bob buy Blackbeard from Carlos and his father, they take him to their *headquarters* where he will stay for the entire audio play series. While not present in the first episode picked for analysis (episode 2, "The Secret of Phantom Lake"), Blackbeard appears in the following (episode 6, "The Mystery of the Talking Skull", time code 35:09). When the three detectives enter their *headquarters*, the *cawing* of the bird can be heard, now a *soundmark* for the boys' trailer. Up to the most recent episodes, Blackbeard is often the only source of information for listeners curious about where the story is taking place. As soon as *cawing* is heard in *TTI*, the place of action is the *headquarters*.

As *soundmarks* are typically sounds of environments, like a neighborhood, that are defined by the local community, this soundscape element is rather rare in both audio plays. The only person to classify *soundmarks* as such is the listener. In addition to listening habits and frequently returning sound objects, social connotation is important to recognizing a *soundmark*. The only example from all 70 episodes is the cry of an *owl* in *TTI* episode 78 ("The Empty Grave"). The occurrence of the *owl* is especially interesting, as it does not appear as a signifier for nighttime – as one might expect – but the graveyard. The *storyteller* introduces the place of action as he tells the listener that "very early the next morning Justus rode to the beach with his bike, where he came to the graveyard of Rocky Beach" (*TTI*, episode 78, time code 09:32–09:50, translated from German). Justus is looking for two graves where his parents might be buried when he runs into his girlfriend, Lys. They are having a conversation, when, in the background and not very loud, the cry of an *owl* can be heard (time code 11:22). While additional *keynote* sounds like *birds chirping* and *leaves rustling* are used as part of the soundscape of various outdoor places of action, the *owl* only appears at the graveyard as a signifier for this particular place of action.

Signal creates sense of place rather implicitly. While *soundmark* directly refers to a specific place of action, *signal* can occur in various locations. It disturbs both characters of audio drama and its listeners, and adds to the description of a place of action. Overall, there are only five different *signals* occurring in *TTI*'s 40 analyzed episodes. Aside from the human sound of a *scream* and the technology sound of a *whip* (see examples below), there is a *SMS tone*, an *alarm clock*, and the sound of *screeching tires* (all *technology* sounds, too).

The first example of a *signal* is a *scream* from *TTI* episode 118 ("The Dark Legacy", track 5, time code 02:35–02:37). Peter and Bob walk toward a trailer while searching for a burglar. Behind the trailer they enter an old garden with ivy and wildflowers. A path leads to an old house without windows. After the description of the scenery by the *storyteller*, *dialogue* of Peter and Bob begins, accompanied by various *keynote* sounds like the *cry of a raven*, the *squeaking of a door*, and the boys' *footsteps* on *groaning floorboards*. As they discover a piece of fur under which something seems to be moving, they get closer and pick it up to find out what lies underneath. Suddenly, a loud scream shocks both the boys and the listener.

Another example is a *whip* occurring in *TTI*'s episode 160 ("Mysterious Messages", track 7, time code 05:54). Justus, Peter, and Bob are in the city archives, where they hope to find the answer to a riddle. A professor and his daughter accompany them. There are some *keynote* sounds like *walking*, *rustling* of paper, and *chirping* of birds outside while the boys search the archives' files. When Bob finally finds what they are looking for they are taken by surprise by the villain and his accomplice. After a battle of words the villain suddenly brings out his whip and cracks it. The boys, the professor and his daughter, and the listener are startled by its sound.

Signals in *Gabriel Burns* are exclusively technology sounds. The lack of nature sounds might be explainable by the focus on the urban environment, but there seems to be no reason that human sounds are not used as *signal* (see the *scream* in *TTI*). The first example in *Gabriel Burns* is the *horn of a car* (episode 6, "Death Machine", time code 01:38). In the beginning of the episode, Danny and Kate, a young couple, are driving through the rain. They are cut off by another car and Danny starts yelling, honking the horn at the driver in the other vehicle. The driver gets out of the car and starts to frighten the couple. He pulls out a gun and forces Danny and Kate to follow him to his vehicle. They drive away together. *Rain* falling both on the car and the streets is one of the dominant sounds of the scene. Together with *rolling thunder* and the sounds of the car (*screeching tires*, *wipers*, *breaks*), *keynote* is drowned by Danny honking the car's *horn*. In contrast to the sound of *screeching tires*, which has the same loudness and intensity, the *horn* is actively integrated into the storyline as a non-verbal element of communication. The *tires* screech incidentally while Danny is talking and driving, and the *horn* cuts into the dialogue: "[*Horn*] What is going on? [*Horn*] Hey, are you crazy? [*Horn*] Come on, man. We're in a hurry" (*Gabriel Burns*, episode 6, time code 01:07–02:58).

The second and third examples of *signals* occur in one single scene. Both the *shots* of a semi-automatic gun and the *explosion* of a letter bomb can be heard in *Gabriel Burns* episode 6, "Death Machine" (time code 06:03; 06:13–06:17). The action takes place in the office of Mr. Bakerman. The secretary is opening the door when a courier delivers a package for Bakerman. She wants to receive it for her boss but the courier insists on delivering it personally. A security man contacts Bakerman on the intercom and Larry Newman, one of Bakerman's employees, appears. The courier is confused because he didn't expect any more people around. Another door opens and Bakerman himself steps into the office. When Bakerman recognizes the courier as one of his major nemeses he pulls a gun and fires at him. The courier collapses but reveals a bomb, which detonates just when Bakerman and his team run for the door. The scene is enriched by several keynote sounds like *footsteps* from all the characters, the opening of a *zipper* when the courier pulls out the package, and the *groaning* of the villain after being shot. The office is represented by *ringing telephones* and the sounds of a *typewriter* and *rustling paper*. The *shots* of Bakerman's semi-automatic gun and the *explosion* caused by the courier's bomb are two *signals* that step out of the *keynote*, as they interrupt the dialogue and express a dramatic turn of the plot.

All five examples from both audio plays show that *signals* always have an impact on the story – as well as the listener. The sirens of an ambulance or church bells ringing belong to an audio play's *keynote* if they do not disturb the characters in their actions. A *scream* or *call* by one of the characters also generally does not count as a *signal*. Most of the time they part of the *dialogue* and have to be rated correspondingly. *Signal* can not only drive a story forward but also add some unexpected twists to the plot.

5.2.3.3 Atmospheric background – keynote

The last soundscape element is *keynote*. It combines human, nature, and technology sounds – sometimes even atmospheric music – into a collage that supports *dialogue* and even *storyteller*. As the cloud in Figure 5.3 indicates, there are numerous sounds in a recording. What seems to be a cacophonous mix can be deconstructed through close listening. Not every listener will recognize each single sound, but the final mix has to be easy to understand and must not confuse the recipient.

> We find that the structure of the sounds and noise must be understandable to the listener because it is his ability to associate sounds with certain types of settings that enables him to complete background noise with his imagination (Wissmann and Zimmermann 2010: 371).

To take a closer look at how the urban environment is constructed through *keynote* in *TTI* and *Gabriel Burns*, sound objects of nine places of action are deconstructed and categorized as *human*, *nature*, and *technology* sounds. As the right table in Figure 5.3 shows, most of the time one category of sounds dominates a place's *keynote*. Sometimes human, nature, and technology sounds are more balanced, while there are also places of action where a category is totally missing. Starting with the most dominant categories, the following examples will show how single sound objects – labeled *h* for *human*, *n* for *nature*, and *t* for *technology* – are composed to create both Rocky Beach and Vancouver.

In both audio plays human sounds are most dominant in the *police station* place of action. During their research in *TTI* episode 70 ("Shots in the Dark") Justus, Peter, and Bob meet with inspector Cotta from the Rocky Beach Police Department in his office (time code 21:13–26:20) to discuss a plot and exchange information. Most *keynote* sounds are rather quiet, as Cotta is *shutting the door* (t) when the three boys arrive. Aside from a *telephone ringing* (t) and the sound of the *groaning floorboard* (t) when people *walk* (h) over it, all other sounds are *human.* There is a temporary *rustling of cloth* (h) whenever somebody moves. From outside the office, the voice of a *police radio* (h) can be heard, as well as some *talking* (h), *coughing* (h), and *walking* (h). The scene is interrupted by Inspector Cotta taking a phone call (time code 23:35). After the second part of the conversation Cotta leaves his office and the scene ends (time code 26:20). There are other examples of the *police station* in *TTI* that contain more *technology* sounds (see episode 94, "The Black Monster", time code 24:30–28:40) like those from *paper* (t), a *felt-tip pen* (t), and the *closing of a suitcase* (t). Overall, human sounds are dominant.

The Police Station in *Gabriel Burns* (episode 1, "The Whisperer", time code 30:46–31:19) offers the same dominance of *human* sounds, but with largely different sound objects. The scene is set in the police station in Dawson, in front of the office of constable Landon. Bakerman and Joyce Kramer are standing in front of the office door when Bakerman *knocks* (h). He tries to open the door, pushing

Uppercase number signifies the number of different places of action, where a sound appears.
Uppercase letter signifies whether a sound is a human (h), nature (n), or technology (t) sound.
Rocky Beach = white Vancouver = black Design and Data (08/2013) by T. Wissmann.

Soundscape Element	Rocky Beach			Vancouver		
	human	nature	technology	human	nature	technology
Keynote	40.81%	10.31%	48.88%	34.33%	12.88%	52.79%
Signal	70.00%		30.00%			100.00%
Soundmark		100.00%				

Place of Action	Rocky Beach			Vancouver		
	human	nature	technology	human	nature	technology
Apartment	41.86%	8.14%	50.00%	40.00%	2.67%	49.33%
Car	31.25%		68.75%	3.70%	12.96%	83.33%
Cellar	33.33%		66.67%	33.33%	25.00%	41.67%
Graveyard	30.00%	40.00%	30.00%	33.33%	50.00%	16.67%
Harbor	33.33%	33.33%	33.33%	28.57%	21.43%	50.00%
Hospital	50.00%		50.00%	30.77%	7.69%	61.54%
Office	35.71%		64.29%	50.00%		50.00%
Police Station	54.55%		45.45%	71.43%		28.57%
Street	36.84%	21.05%	42.98%	49.06%	20.75%	30.19%

Figure 5.3 ***Gabriel Burns* and *TTI* – Sound Cloud**

the *handle* (t). Mad *laughing* (h) and *murmuring* (h) can be heard coming from inside. Bakerman tries to start a conversation with the obviously mad constable. He takes a few *steps* (h) closer to the door and talks to the man. *Cloth rustling* (h) can be heard. Because Bakerman only receives weird answers to his questions, both he and Joyce Kramer *walk* (h) away from the office.

Nature sounds can be found mainly outdoors. Places of action like *graveyard* consist of a high number of them. Even if a particular story does not play out in closed buildings or vehicles, *nature* sounds generally never dominate the soundscape of a scene. The only exception within the urban environment is the *graveyard.* In *TTI* episode 106 ("Headless Man", time code 49:10–53:02), Justus, Peter, and Bob are hiding behind a hibiscus tree to witness a funeral. The *storyteller* talks over the scene and introduces the place of action. It is *raining* (n), *thunder* (n) can be heard, and a *church bell* (t) is ringing in the background. *Heavy raindrops* (n) appear closer to the boys, as *birds hop* (n) through the bushes, which are represented by some *rustling* (n). *Leafs of trees* (n) move in the *wind* (n) as Justus receives an *SMS* (t) and the boys discuss its content as well as the episode's plot. In addition to the *deep calling of a bird* (n) close to the place where the *coffin is lowered* (t) into the ground and a priest is saying a *prayer* (h), high *chirping* (n) from another bird appears close to the *rustling hibiscus bush* (n) where the three detectives are hiding. *Wind* (n) masks their voices a little as the funeral ceremony comes to an end. *Children's voices* (h) can be heard in the background, as well as a sound like the *closing of a car door* (t) and the *reverse gear of a car* (t). People are probably leaving the graveyard. The scene ends with a very loud *clap of thunder* (n) and the beginning of *soundframe.*

In the *graveyard* scene in *Gabriel Burns* episode 17 ("What is Life?", time code 04:06–06:07) the characters do not watch but participate in a funeral. Aside from the *priest* speaking a prayer (h), the dominant *keynotes* are *rain* (n), *wind* (n), and a *church bell* (t). Stephen stays at the coffin, which finally is *lowered* (t). He grabs a shovel and throws *dirt* (n) into the hole. *Dripping water* (n) indicates that Stephen is getting wet. The only *human* sound from people standing at the grave is that of their *shoes changing position* (h) as they listen to the priest. Weeping, talking, and people's clothes rustling are not present. After the funeral sounds from people *walking* (h) away can be heard. At that time the scene ends and *soundframe* – eerie, churchy music – begins (time code 06:07).

It is no wonder that technology sounds are used most to evoke the city in audio drama. As shown above for the urban environment (see Chapters 3 and 4), sounds from this category can be heard most when walking through a city's streets. The city in audio drama is no exception. In both *TTI* and *Gabriel Burns* technology sounds dominate *keynote.* In particular, places of action with a lot of technical equipment and machinery allow for almost no other sound objects – mainly because of the masking effect of generally louder technology sounds. The *car* is the best example of this. In *TTI* episode 94 ("The Black Monster") (time code 02:04–02:45) the three detectives want to celebrate Peter's birthday. Justus and Bob lead him to the car. The sounds of *opening doors* (t) and *clothes* (h) on the seats indicate that they all get ready to leave. *Doors close* (t), and the *key* (t) is put into the ignition. One of the boys starts the *engine* (t), *accelerates* (t), and *changes gear* (t). Peter has no idea where they are heading. To give him an idea of what might be happening later, the boys turn on the *radio* (t). The music from the sound system transforms from *keynote* to *soundframe*, and as the *storyteller* starts to talk the scene ends (time code 02:45).

The *car* in *Gabriel Burns* episode 2 ("The Brood") (time code 04:05–06:32) is Stephen's cab. Before he becomes a full-time employee for Bakerman and starts to fight the mysterious Others, he works as a cab driver in Vancouver. The scene begins with the sound of the *car engine* (t). *Radio static* (t) and some *talking* (h) within indicate that the place of action is a cab. This becomes obvious when the cab operator *calls* (h) Stephen to tell him about a new customer to pick up. He *accelerates* (t) and *shifts gear* (t). While looking for the right address and probably turning left and right several times, Stephen's *clothes* (h) can be heard whenever they touch the seat or steering wheel. *Keynote* is ducked for the *storyteller* giving more information on the scene until Stephen is close to the address. The cab radio (h) is still on when the *wheels stop* (t) on a grainy street and Stephen waits with the *engine running* (t). Suddenly the *door opens* (t) and Bakerman gets in the cab. *Keys* (t) jangle. The *door* (t) is shut and the men start having a conversation. *Cars pass by* (t), building the background of their conversation. After a short interlude – during which Stephen remembers his childhood (time code 06:01–06:06) – the scene ends (time code 06:32).

Some places of action have a very balanced *keynote.* All three categories of human, nature, and technology sounds are represented more or less equally, as the following examples show. In *TTI* episode 152 ("Skateboard Fever") (time code 43:57–48:21) Peter is driving to the harbor to look for a homeless man he knows. When he arrives *seagulls* (n) and people *talking* (h) can be heard. While he looks for the man, Burt Young, an old friend, *steps* (h) closer from behind and addresses Peter. During the conversation about espionage, rival countries, and a secret project, the *keynote* employs a lot of different sounds to make the harbor present for the listener. Both *bikes* (t) and *motorbikes* (t) pass by the two characters, and there are *voices* (h), *shouting* (h), and *laughing* (h) from people on site. *Chains* (t), maybe from a gate, can be heard, as well as compressed air from a *bus* (t) leaving its stop, *chugging* (t) – probably from a small boat – and a ship's *horn* (t). *Wind* (n) and *water* (n) are nature sounds that, combined with *seagulls* (n), add to the natural atmosphere of place. There seems to be at least one restaurant not too far away from the point where Peter and Burt are standing. *Money dropping* (t) and the sound of *dishes* (t) add to the *keynote.* The scene ends with Peter leaving the harbor and *soundframe* setting in (time code 48:21).

In *Gabriel Burns* episode 28 ("In the Circle of Trust") (time code 05:13–7:07) Stephen and Larry enter a cellar. Their *shoes walking* (h) down the stairs can be heard. Larry comments on how cold it is and the sound of a *zipper* (h) indicates that he is closing his jacket. Larry tries to turn on his *flashlight* (t). They take a look around. Stephen says that he can feel the magical presence of the Others.

The *dialogue* is presented with strong reverb, so that the *echo* (h) of the voices becomes part of *keynote*. As Stephen and Larry continue walking, soft *weeping* (h) can be heard. A little girl is sitting in the cellar, which now appears to be more like a huge cave. *Dust* (n) on the ground can be heard while the men come closer. *Water* (n) is dripping from the walls, and *wind* (n) starts to blow. Stephen is drawn into the energies of the Others and the scene fades out. During the whole scene, music is used in the *keynote's* background to support the atmosphere and add a mysterious touch to the recording. As stated above, sometimes audio drama has to portray the sounds of sources that do not exist in our environment – like magic. Abstract concepts like mystery and horror also have to be evoked. Keynote music is a good way to do so. In addition to the (common) sound of *thunder* representing magical eruptions in media, music generates responses in the listener that help to evoke the abnormal and other. The *Gabriel Burns* scene in the cellar is a good example of this, where reverberation is also used to express that the place of action is somehow removed from reality.

As the above examples show:

> the listener's imagination is both inspired and actively engaged simultaneously by the range of acoustic signals and information provided in addition to language (Wissmann and Zimmermann 2010: 379).

To make sense of what the listener hears it is important that quiet sounds like *keynote* can still be perceived. Thus, the listening environment plays a huge role, as does the stereo on which the recording is played back. Producers of audio plays originally created for children, in particular, have to take this into account. Regardless of playback device and environmental sound the story of an audio play has to be understandable and its world has to be transported to the listener. If the listener does not understand what to make of single sound objects, those objects are more distractions than tools for constructing places of action.

> Therefore, the structure of the sounds and noise must be understandable to the listeners (Wissmann and Zimmermann 2010: 379).

To guarantee the greatest success possible in reaching a large number of listeners, sound objects have to be easy to understand and listeners must be able to assign them to respective sources. *Keynote* sounds in particular are difficult to deconstruct, just because of their low volume. One way to ensure a good listening experience is to use sounds every listener can easily understand. Reoccurring sounds may help the listener follow the story. If every door in an audio play sounds different, then this door will be harder to recognize. *TTI* episodes clearly offer a much simpler solution. There may be the same number of sound objects as there are in a comparable *Gabriel Burns* production, but there is less variety.

Whether one or the other solution produces a better story is up to the listener. The plot is mainly transported through *storyteller* and *dialogue*, so in the end additional soundscape elements like *keynote*, *signal*, and *soundmark*, and even *soundframe* seem to be less important. But to evoke the city in various places of action is crucial to evoking the atmosphere of both plot and genre. Places like the *cellar* in Vancouver appear to be mysterious, eerie, and different from our familiar surroundings. It is not a cellar you would enter to pick up a bottle of wine. All soundscape elements help manifest the special atmosphere that supports the story. The listener gets an idea of how Vancouver and Rocky Beach look – even though there is nothing to see. The city is evoked through sound and, importantly, the listener's imagination. As the above examples show, not every object that exists in a place of action is represented through sound. The listener has to fill in blanks to create a full image. This way, the listener does not get overwhelmed by a cacophonous soundtrack, and it is possible to customize the image of a city based on one's own stream of consciousness.

Chapter 6
Sound Is the City

6.1 The Missing Sense

It seems natural to keep an ear out for sound. But explicitly listening to sound within the urban landscape is something we are not really used to. Visual elements are far more dominant and make a greater impact on our perception of the world. They are eye-catching, so to speak. When you think of any city, your thoughts first go to skylines, streets, and landmarks, not sound. But the acoustic element, just like the visual, plays its part in co-shaping what makes a city special and unique. When *sense of place* is discussed, the focus primarily lies on visual environmental aspects (Banerjee and Southworth 1990, Tuan 1975) and social values (Cresswell 1996, Halpenny 2010), not sound (except, Boland 2010, Johansson and Bell 2009b, Rodaway 1994). What ought to be true for media, where zero sound is avoided at almost any price (Figgis 2003), seems to be general practice in sense of place studies. As the discussion has shown, sound too often is tuned out.

First contact with the sonic environment really begins with focusing on sound. Schafer's experiments with students are, in my opinion, still the most impressive way to confront the urban dweller with the inevitability and integral presence of sound (Schafer 1994). Everyone who cares about his/her own environment and the effects it can have on his/her personal view and construction of the world should at least once perform his/her daily routine without speaking. If this means keeping quiet while sitting at the home office's desk, the effect might not be as bold. But walking around campus and buying a snack without having any conversations can be a real challenge. The aim is not to engage in anti-social behavior but to focus solely on sound. Active listening is key to any sound study. While doing so, visual input distracts listening and lets us shift our focus away from sound to the more familiar visual impressions. It is not so easy to mainly listen when one's eyes are open.

Starting to hear what occurs in the city and being sensitized to urban sound opens up the question, 'What in the world do we hear' (Gaver 1993). Acoustic ecology, again, helps deconstruct the cacophony. Generally, it does not really matter how the soundscape is split up into single sound events. Realizing that deconstruction has to be done to know how to cope with surrounding sounds is of major importance. As shown earlier, there are numerous ways to categorize and classify sound events (Papadimitriou et al. 2009, Servigne et al. 1999, Yu and Kang 2010). Schafer's proposed trichotomy of *human*, *nature*, and *electric/mechanical* sounds (which I summarized as *technology* sounds) is one way to organize urban sounds – and those in media. It shows that differentiating between those three categories of sound events is helpful for learning what kind of sounds have more positive or negative impacts on the urban dweller's life. Chapter 2 presented numerous examples of urban sounds that effect us in one or the other way. Sounds like a motorbike's engine or live music from an open-air festival might be pleasant for some people while others classify it as noise. The physiognomy of the human ear restricts what can be heard. High decibel levels cause pain and permanent hearing loss while sounds with low levels may not be heard at all. Sounds, especially music, can generate states of joy, even ecstasy, regardless of volume (Becker 1994, Brazeal 2003, Jankowsky 2007). Tuning out environmental sounds and replacing them with music from portable music players (Bull 2000) can even personalize the soundscape. Why is there such a need to avoid urban sounds? Is the city simply too noisy? Soundwalks, sound event recording, and especially *sound point recording* help to analyze soundscape composition in the urban environment. Knowing what sounds occur where (and when) within the city is crucial for dealing with related issues (see 6.3). The methodology in Chapter 3 proposed a new way to learn more about sound sources and their locations. In five selected research areas in Austin, TX, human, nature, and technology sounds were recorded on soundwalks that helped deconstruct the soundscape. These soundwalks showed that downtown areas predominantly consist of loud technology sounds from traffic, construction, and music, while urban green spaces and single-family neighborhoods are quieter. In order to be heard, nature sounds, in particular, like the chirping of birds and the rustling of leaves, require a silent environment where they are not masked by passing vehicles or turned-up loudspeakers. The analysis of all sounds occurring in each research area allowed the researcher to perform sound point recordings, for which a single location was chosen for identifying representative sound events of the respective area. Sound point recording delivers place-based information about the soundscape. Never before has sound been set into context of place in a way that enables us to learn more about the impact sound has on sense of place.

How do urban areas appear to us? If you are wandering around Hyde Park in Austin, TX, described in section 3.4.3, you're experiencing a neighborhood created in the late nineteenth century. The former whites-only suburb has been transformed since then. Single-family houses and bungalows from the 1920s and 1930s are typical for the area, as are numerous green spaces, gardens, and trees. Occasionally cars pass by and you can meet walkers, joggers, and bikers. The image just described is at best half-finished, as there is no mention of sounds and other sensuous impressions. In addition to houses, streets, and cars there are sounds, too – the singing of birds, children laughing in backyards, lawn mower engines humming around the corner: Without sound we don't get the complete impression. Just like visual impressions, sound is part of sense of place and must not be ignored. If Tuan's description of place as "a center of felt value" (Tuan 1977: 138) is true, do we value sound so little that visual aspects occur almost exclusively when we think about place? I argue that we do think about sound when we think about place. Yet our evaluation of place is generally based unconsciously on sound. Why this is the case is hard to determine.

In the case of unwanted sounds, it may be possible that we got so accustomed to living in a mostly loud (often labeled as "noisy") environment that we take bad sound quality for granted. Maybe we are under the impression that we are "living in the city, what do you expect?" (Leasure 2011). Only in extreme situations, like the one described in 4.2.3 using the example of air traffic noise, do residents seems to be willing to tackle sound issues. Austin modified its 3–1–1 hotline so that residents could file noise complaints. But do we ever hear people say, "This is a nice-sounding neighborhood"? Places of silence are the opposite of noisy areas. Buildings like the sanctuary of a church (see 4.5) are places where we also become aware that sound adds to sense of place. But sound adds to sense of place in all places. Sound should not only matter in churches or the approach path of an airport. Previous discussion has shown that sound is always part of place and co-shapes the way place is created. Every listener will describe sound in a different way; individual stream of consciousness defines how we perceive and experience the urban soundscape. Sense of place, as the term indicates, has to do with sensing. Individual and collective experience of place defines the world we live in. Places like live music venues, neighborhoods, and cities as a whole are connoted with various labels, ideas, and prejudices. Sound is always a part of it.

As stated above, sound is mostly missing in discussions on sense of place. Whenever it occurs, it is described as being *in* place. Music genres that have shaped the image of cities like New Orleans, Nashville, and Detroit contribute to each city's sense of place (Che 2009, Pesses 2009). In live-music venues artists perform and thus co-shape sense of place. Language, especially dialects that can be linked to place, like Liverpool's *scouse*, identifies locals with their neighborhood (Boland 2010). Radio shows and even entire stations are able to unite social groups with the help of music (Lindenbaum 2009). But sound is never recognized as part of place. Music takes place *in* live-music venues, sounds are created *in* a certain city, and a dialect is spoken *in* an area. But what do all these places – buildings, neighborhoods, cities – consist of? It is not only what we see that makes a place a place. What more is place than all its perceptible elements? Thus, we have to stop thinking of sound only as being *in* place. Sound *is* place. Sound *is* the city.

6.2 Place-Based Sound

Understanding sound as place shifts the focus of studies in the humanities. Acoustic ecology's approach, to divide the soundscape into *keynote, signal,* and *soundmark,* already carries the place-based character of sound in it, at least in one part of the trichotomy. Keynote, as background sound of place, is considered to be *in* place. Traffic in the streets, human voices in a park, and birds singing in the trees all are taken as sound sources that are not synonymous with place itself. Soundmarks are different. They are described as *community sounds* (Truax 1978) that are linked to cultural heritage and social norms. In addition to a soundmark's constructive component, it is always linked to place. The bells of Big Ben have been named as an acoustic landmark for the city of London, as has the singing of canary birds for the Alfama neighborhood of Lisbon, Portugal (see 3.4.2 and 3.4.1). While acoustic ecology focuses more on the fact that soundmark is connoted by the listener, *geographies of urban sound* set it into spatial context. Soundmark, as sound, not only occurs in a certain place; it co-defines that place. The place-based character of sound becomes even more obvious with the presented audio media extension of the soundscape (Wissmann and Zimmermann 2010). Sound is the only way in which space and places can be created in audio drama. Without sound there would be no *Rocky Beach* or *Vancouver.* Accordingly, places of action are audible because of *storyteller's* descriptions, characters' *dialogue*, and interluding *soundframes.* As 5.2 has shown, every element of an audio media soundscape evokes sense of place that is formed by the listener, including her/his own stream of consciousness. In addition to acoustic ecology, there are numerous fields where *geographies of urban sound* have the potential to change the research focus and ask slightly different questions.

In architecture, where visual expression is generally more important that the acoustic qualities of materials, sound is often left out of the construction process. The discussion about missing noise insulation in apartments close to live music venues mentioned in 4.2.2 shows that economic and other issues might stand in the way of proper sound design (Leasure 2011). Opera houses, theaters, and city halls are exceptions, where sound wave propagation is considered. While architects focus on the interior sound design of such places, the shape and material of facades and the location of other buildings are largely ignored (Cavanaugh, Tocci, and Wilkes 2010). Whenever effects of sound are integrated into the construction process we can speak of *aural architecture* (Blesser and Salter 2007). We need to stop thinking of sound as being attached to place and starting thinking of sound *as* place. Acoustics belong to sense of place just like concrete and stone do. Thus, labeling architecture as specifically *aural* should not be necessary. Who would speak of *visual architecture* to describe a building?

Psychology, for example, focuses more on the individual and social impact of sound. Sound is described as occurring in places and co-shaping the image of those places (see Kim and Rowland). While the urban soundscape constantly gets louder and more cacophonous due to urbanization and new sound sources (see Fortuna 2001), many people decide to mask environmental sounds with personal sound from portable music players like iPods and smartphones. In subsection 2.5.2 we discussed the influence of those devices on transforming not only the acoustic surroundings of the listener but the space itself. In my opinion, the transformation of public space into multiple private spaces should be one key concern of sociological urban studies (see Bull 2007, Simun 2009). Realizing that sense of place is based on individual perception and one's own stream of consciousness that the listener actively builds, modifies, and adjusts, the question about how this new place and its sense are shaped has to be asked. Discussion shows that sense of place is never ontologically inherent to a certain location, neighborhood, or city. Even without a personalized soundtrack, every urban dweller perceives and constructs place differently. The focus on sound and *iPod culture* (Bull 2007) highlights this thesis and – better than a purely visual analysis could – enables us to find out more about sense of place creation.

Art installations are one of the best ways to realize how sense of place can be altered by sound. The example of *Anhalter Bahnhof* described in 1.4 shows that adding an artificial soundscape to place changes the way it is perceived (see Fontana 1987). Sound appears to be brought to a place where it expands its potential and affects every listener. *Geographies of urban sound* want to focus on the place-based character of sound. While sound can be thought of as adding to place, it can also be understood as an integral part of it. Thus, it is not only the recorded sound of trains that changes the soundscape of *Anhalter Bahnhof*, but place itself. The place perception and experience of every visitor to such an installation are permanently changed because the stream of consciousness is altered by new input. Thus, even temporary art installations modify sense of place forever – or at least until the recipient processes new sensory input in a way that makes the current place image irrelevant.

There may be a lack of *museums of lost sounds* (Schafer 1994) but with *sound point recording* we get the chance to record typical sound events in an area. Through the deconstruction and analysis of sound objects – they are no longer sound events as soon as they are pulled out of their environmental context (Truax 1978) – representative sounds can be detected. Leaving decibel measurements behind this method strengthens the social perspective. Noise ordinances, including those with maximum values for sound depending on the time of day, might be an expression of intersubjective needs and common sense. But they also change the subject from individual perception to quantitative regulations. Sound levels tell us nothing about the source causing the sound. And without focusing on the wide variety of sound events from nature, human, and technology, questions about how those sounds are perceived, experienced, and taken can never be answered.

6.3 Geographies of Urban Sound

The social geographic perspective drives the realization of the place-based character of sound. Without the spatial focus it is not possible to integrate sound to sense of place, as described in previous chapters. While this shift of perspective can be applied to other fields in the humanities, human geography, especially social geography, might benefit from another realization: Sound is not only music. This seems rather obvious – remember Schafer describing music as the organized form of sound (1994) – while current geographical studies almost exclusively focus on music (see 1.3) (except a few publications like Jazeel 2005, Matless 2005, Rodaway 1994) or language (Boland 2010). Johansson and Bell's collection, *Sound, Society, and the Geography of Popular Music* (2009a) is a good example. While the title of the book explicitly names *sound*, contributed papers deal with a specific form of sound: music. This focus on music can also be found in discussions at various annual meetings, such as those of the *Association of American Geographers*. Labeled as *media geography*, studies on sound are presented in media sessions, all of which primarily focus on music, not its broader sound context (at least in meetings until 2013). We do not have to limit ourselves in such a way. Geographic research should expand and deal not only with music.

The examples in Chapter 4 – one of which actually deals with music – show that there are numerous questions to investigate if we want to understand the human-environmental relationship regarding sound. The basic observation is that in social contexts, measuring decibel levels makes no sense. Air traffic noise in the landing path of Frankfurt Airport can be described vis à vis its loudness. Aircraft altitudes and weather conditions determine how loud turbines are at landing. How sound waves are processed is up to the receiver, i.e., the urban dweller who hears a plane flying by. Sensitization to air traffic sound and individual perception are decisive when determining how that sound is evaluated. Whether the sound from aircrafts is considered wanted or unwanted, and whether it is rated loud, medium, or even low is entirely up to the listener. In the context of former experiences stored in the stream of consciousness labels like *noise* are applied to air traffic sound or not. The same goes for live music of different genres and during various occasions. While the DJ at a neighbor's might cause major annoyance to someone who wants to sleep, the same DJ might be valued as the guarantee of a successful wedding reception by the same person at a different time. Just as individuals react differently to sound, businesses might also be affected by the environmental soundscape. Is noisy Times Square in New York City a good place for your office? This might depend on the type of business you are running. Creative think tanks might be annoyed by the sounds of traffic and tourists. For a news agency this might be the exact place to ensure employees have the feeling of having their fingers on the pulse of current affairs. In this context, sound is not something that can simply be measured.

The technological use of sound has a tremendous impact on society. Due to the rise of portable music players and personalized listening, public space is transforming into multiple private environments. Audio-guided tours cut the link between the local population and visiting tourists, who, in turn, are relying on the storyteller in digital recordings rather than a live tour guide. Thus, one might argue that the sound of audio tours supports a less personal atmosphere while at the same time allowing visitors to experience new places. But it can also be understood as promoting the personal experience. Conversation with a prerecorded tour guide is not possible, but the choice to alter the soundscape is an expression of extreme personalization. Concerning the urban environment, headphones and artificial soundtracks raise a number of questions for further research, while sound in media opens up a whole new field concerning the latter. Audio drama consists of sound only. As part of popular culture media, audio drama alters perception of the city and any other place. Children's audio plays co-shape the listener's ideas and expectations of the world at an early age. The interplay of music, language, and environmental sounds temporarily replaces otherwise predominant visual input and supports active listening. The images evoked through audio media like audio-guided tours and audio drama change the interaction between the listener and her/his environment. The transformation from public to publicly private space again demonstrates how powerful the impact of sound on sense of place is. Individual perception is altered not only by environmental sound but also sound in media.

Sound coming from the personal stereo often connects with visual and other sensory impressions the listener perceives from the environment. To evoke a specific image of a city, tourist agencies and city marketing try to "enter" the playlist of visitors via a suitable soundtrack. As discussed in 4.3, Austin, Texas, annually produces a music album that is supposed to represent the city's image acoustically. Genres like blues, rock, and country are used to transport the image of a culturally vibrant metropolis of typical US-American music that cultivates its roots, going back to artists like *Willie Nelson* and *Stevie Ray Vaughan.* Instead of putting effort into marketing local landmarks, *soundmarks* are used to co-shape people's sense of place. My analysis showed that music as a marketing strategy can influence not only visitors but also locals as prominent news media cite the sound-based image and thus affect the urban dweller. From a Marxist perspective power structures are expressed through music marketing – e.g., certain genres are less represented and stereotypes are used to sharpen a city's image. Thus *geographies of urban sound* are also about preference and oppression, about simplifying and banalizing. Sound can be used to define a city, its (medial) image, and the personally perceived sense of place.

Whenever sound is missing in the urban environment, it attracts attention. Quiet spaces are understood as helping the urban dweller to relax and find some peace within the daily hustle and bustle. Too much sound is generally thought to negatively affect a person's health. What does this mean for urban planning? Do we have to spend even more attention not only on public green spaces but also public quiet places? Or have we already solved the problem with our portable personal stereos? Music from an iPod, in addition to affecting perception of the urban environment, has the potential to mask unwanted sounds and tune out incriminating noise. Maybe it is in the sounds of a playlist that we can find silence, just like the sounds of the services in church aim at creating an environment for silence (see 4.5) (Barr 2012). It is also important to realize that sometimes it is okay if sound is missing. Urban spaces should not follow the dictates of (visual) media that state that you cannot have zero sound (Figgis 2003). We can and should allow ourselves to completely tune out the urban soundscape, at least sometimes. Medical studies have shown that listening to the radio at work interferes with performance, even though it might not seem to be disturbing to the individual. *Geographies of urban sound* have to include discussions about silence in the city and its potential benefits. At the same time fear of unwanted silence has to be addressed. Individual restrictions and prejudices against silent areas lead to places being avoided by people (cf. Valentine 1989). Silence can trigger a feeling of being less safe in a city because of missing (positively connoted) sounds. Urban planning should consider these possible sound effects.

Finally, *geographies of urban sound* are important because they promote an anti-essentialist view of the world. It is the purpose of other scientific disciplines to find out and work with an existing and taken-for-granted urban soundscape. High decibel levels from inner-city highways may result in the installation of noise-protection walls. Live music concerts can induce the city to issue special permits and prohibitions. From a social perspective the source is less relevant. It is the receiver, i.e., the urban dweller, who matters. Based on radical constructivist and phenomenological views, it is the listener who actively perceives and processes the soundscape. Consequences in acting and behaving are investigated. It is the subjective character of the (perceived) soundscape that causes evaluations like *wanted* or *unwanted*, and creates labels like *noise*. Thus, we should not ask how it sounds somewhere, or if there is a typical sound for cities like Austin, Frankfurt, London, and Lisbon. Instead, it would be far more valuable to find out about how we, the listeners, are shaped by sound that evokes specific kinds of associations, causes actions, and triggers memories. *Geographies of urban sound* do not provide essentialist cultural assumptions but speak to the influence of environmental sound on the urban dweller and the impressions it evokes.

6.4 Breaking the Sound Barrier

In the end, sound is not the only sense that geography should focus on. This book does not want to create the impression that sound should be superior to other senses when it comes to shaping our view of the world. The expression 'sound is the city' does not mean that there are no other elements co-constituting the urban environment. *Geographies of urban sound* focuses on the contribution of sound to shaping our (active construction of the) world, with 'contribution' being the most important term. Most theories and models that try to explain the urban social environment are based on visual elements. Concepts like Gentrification – a process of enhancing a neighborhood through the influx of higher-income residents, accompanied by the improvement of local buildings – describe social change and structural enhancements. What about sound? Is sound insulation getting better, is there less violent talk in the streets, and do enhancements include less noise disturbance due to measures like traffic calming? *Geographies of urban sound* wants to sensitize us to think outside the (visual) box and ask about the other influences we are subjected to.

What about our other senses? There are at least three others – taste, smell, and touch – that are worth looking into. Considering all senses in a combined description of the *urban atmosphere* could be one possible way to address the issue (cf. Hasse 2012). For in-depth basic investigation I would encourage research similar to the *geographies of urban sound's* focusing on a single sense like smell, for example. One could try to modify this whole text and replace *sound* with *smell*. What could be said about wanted and unwanted smells? Is smell, like sound, something referring to our own stream of consciousness where familiar smells exist and alien ones distract and worry us? The same thought experiment could be used for the sense of taste. We all have our favorite dishes, which tell us about ourselves and where they were produced. As a quite popular field of study, geographies of food can focus on individual perception and social connotation just like *geographies of urban sound* do with the sense of hearing. It seems to *make sense* to tackle all sensory input that co-shapes our image of the world. As social sciences like human geography want to learn out about human actions and relevant variables, it would be literally short-sighted to ignore every other input aside from the visual.

We have to turn away from the dictates of the visual. This might be hard in a world where the visual input of the environment is even enhanced by media like HD television and the almost ubiquitous use of computers. But media also bears a chance to realize the potential of sound. Movies and TV series come with rich soundtrack that support the atmosphere of the story and sometimes even lead it. As discussed in Chapter 5, sound in this context is mostly recognized when it is missing. Until now, other senses are not integrated in (visual) media. But does that mean that they do not matter? If we think about watching a scene from 2009's *Julie & Julia*, wherein Amy Adams' character, Julie Powell, is cooking the recipes from Julia Child's famous cookbook *Mastering the Art of French Cooking*, you can almost taste the deliciously looking *boeuf bourguignon*. The 2006 movie *Perfume: The Story of a Murderer* – based on Patrick Süskind's classic novel – lets the audience smell how the most bewitching scents and perfumes are made out of the fragrant skin of young Parisians. And when Titanic collides with an iceberg in James Cameron's 1997 movie of the same name, you can almost feel the touch of cold water creeping around the horrified passengers of the sinking luxury liner. I would argue that taste, smell, and touch matter too in media reception. As the analysis of audio drama shows, sound often does not have to be recorded to be heard. The listener adds missing sounds intellectually based on his/her own stream of consciousness. Generalization and simplification of the production stimulate not only the listener's imagination but also his/her ability to complete the image. Although taste, smell, and touch don't occur in media productions they could very well play an important part in co-shaping the image of a medial world. Felt sensory impression would prove our active, individual construction of the world. What we sense can be stimulated by our (media) environment, but, in the end, what we taste, smell, touch, hear, and see depends on us alone. Other senses aside from seeing might be less eye-catching, but their potential influence on perception makes them no less important to investigate.

Deciding to keep quiet for a while and do Schafer's experiment of a day without speaking is a good start to sensitize ourselves to the surrounding soundscape. We have to consider all the senses when it comes to investigating human life in the urban environment and everywhere else. We have to shift away from the dominance of the visual. Only then will it be possible to recognize the various sensory inputs influencing how we perceive the world. In the end, it is the individual who constructs the world based on his/her own stream of consciousness.

Learning about the *geographies of urban sound* means learning about individual perception and active creation. In my opinion, it would be negligent and limiting to look at the sense of sight exclusively. Especially in the urban environment we are confronted with a wide variety of stimuli, all of which influence the way we live and cope with our surroundings. I really hope that this enquiry into the *geographies of urban sound* will not die away but echo in an intensified discussion. Sound is simply too loud to be ignored any longer.

Bibliography

Acheson, K. 2008. Silence as gesture: Rethinking the nature of communicative silences. *Communication Theory*, 18(4), 535–55.

Adams, J.L. 1994. Resonance of Place. *The North American Review*, 279(1), 8–18.

Adams, M., Cox, T., Moore, G., Croxford, B., Refaee, M., and Sharples, S. 2006. Sustainable soundscapes: Noise policy and the urban experience. *Urban Studies*, 43(13), 2385–98.

Adams, P.C. 2009. *Geographies of Media and Communication.* Hoboken, NJ: John Wiley & Sons.

Addyman, D. 2010. Phenomenology "less the rosy hue": Beckett and the Philosophy of Place. *Journal of Modern Literature*, 33(4), 112–28.

Aglioti, S.M. and Pazzaglia, M. 2011. Sounds and scents in (social) action. *Trends in Cognitive Science*, 15(2), 47–55.

Aitken, S.C. and Zonn, L.E. 1994. *Place, Power, Situation, and Spectacle: A Geography of Film.* Lanham, MD: Rowman & Littlefield.

Akstinat, B. 2008. *Das ABC der drei Fragezeichen. Das inoffizielle, unabhängige 3 Fragezeichen Fan-Lexikon vo A bis Z.* Hannover: Humboldt.

Alam, S.M., Eang, L.S., Tan, A., and Tiong, T.S. 2010. An investigation of community noise in high-rise residential environments. *Journal of the Acoustical Society of America*, 127(6), 3511–18.

Alderman, D.H. 2009. Writing on the Graceland Wall: On the importance of authorship in pilgrimage landscapes, in *Sound, Society, and the Geography of Popular Music*, edited by T.L. Bell and O. Johansson. Farnham and Burlington, VT: Ashgate, 53–65.

Alliance for Audited Media, n.d. *Top Media Outlets: Newspapers, Blogs, Consumer Magazines, Broadcast, Websites & Social Networks.* [Online]. Available at: www.burrellesluce.com/topmedia2013 [accessed: 28 May 2013].

Alpert, J.I. and Alpert, M.I. 1989. Background music as an influence in consumer mood and advertising responses. *Advances in Consumer Research*, 16(1), 485–91.

Anderson, B. 2002. A principle of hope: Recorded music, listening practices and the immanence of utopia. *Geografiska Annaler: Series B, Human Geography*, 84(3–4), 211–27.

Arana, M., Martín, R.S., Nagore, I., and Pérez, D. 2011. What precision in the Digital Terrain Model is required for noise mapping? *Applied Acoustics*, 72(8), 522–6.

Arkette, S. 2004. Sounds like city. *Theory, Culture & Society*, 21(1), 159–68.

Atkinson, R. 2007. Ecology of sound: The sonic order of urban space. *Urban Studies*, 44(10), 1905–17.

Atkinson, R. 2011. Ears have walls. *Aether*, 7, 12–26.
Austin Museum of Art, 2010. *Chris Jordan Audio Guide: Main Statistic Report, May 21 to August 15.* Austin, TX: Austin Museum of Art.
Austin Music Commission, 2012. *Austin Music Commission Minutes: Regular Meeting. Monday, February 6, 2012, 6pm.* Austin, TX: City of Austin.
Austin Music People, n.d. *Why AMP?* [Online]. Available at: http://austinmusicpeople.org/about/ [accessed: 27 May 2013].
Bachtin, M.M. 2008. *Chronotopos.* Frankfurt am Main: Suhrkamp.
Baier, H. 2009. Die Vertreibung der Sinne. Klangräume: Rufen und Hören in der verstehenden Soziologie. *Cultura: International Journal of Philosophy of Culture and Axiology*, 6(2), 7–45.
Balzer, M. 2012. Recorded Interview with Author, 2 June 2012.
Banerjee, T. and Southworth, M. 1990. *City Sense and City Design: Writing and Projects of Kevin Lynch.* Cambridge, MA: The MIT Press.
Baranzini, A. and Ramirez, J.V. 2005. Paying for quietness: The impact of noise on Geneva rents. *Urban Studies*, 42(4), 633–46.
Barnes, M. 2011. A pacesetter in the world of concert promotion. *Austin American Statesman*, 6 November, F1.
Barr, D.G. 2012. Recorded interview with author, 4 January 2012.
Barrigón, J.M., Escobar, V.G., Gozalo, G.R., and Vílchez-Gómez, R. 2010. Possible relation of noise levels in streets to the population of the municipalities in which they are located. *Journal of the Acoustical Society of America*, 128(2), EL86–EL92.
Becker, J. 1994. Music and trance. *Leonardo Music Journal*, 4, 41–51.
Berglund, B., Lindvall, T., and Schwela, D. 1999. *Guidelines for Community Noise.* Geneva: World Health Organization.
Berkeley, G. 2009. *Principles of Human Knowledge and Three Dialogues.* New York: Oxford University Press.
Beurmann, A. 2011. *Die drei ???, die Hörspielkönigin und vieles mehr.* Hildesheim: Olms.
Bijsterveld, K. and van Dijck, J. 2009. *Sound Souvenirs: Audio Technologies, Memory and Cultural Practices.* Amsterdam: Amsterdam University Press.
Billings, W.L. 1990. *Effects of Store Atmosphere on Shopping Behavior.* Bloomington, IL: Illinois Wesleyan University.
Blanco, J.C. and Flindell, I. 2011. Property prices in urban areas affected by road traffic noise. *Applied Acoustics*, 72(4), 133–41.
Blanton Museum of Art, 2011. *About Face: Cell Phone Tour Survey. April 30 to September 4.* Austin, TX: Blanton Museum of Art.
Blauert, J. 2001. *Spatial Hearing: The Psychophysics of Human Sound Localization.* Cambridge, MA: The MIT Press.
Blesser, B. and Salter, L.R. 2007. *Spaces Speak, Are You Listening? Experiencing Aural Architecture.* Cambridge, MA: The MIT Press.
Bloch, E. 1965. *Literarische Aufsätze. Gesamtausgabe.* Frankfurt am Main: Suhrkamp.

Boland, P. 2010. Sonic geography, place and race in the formation of local identity: Liverpool and scousers. *Geografiska Annaler: Series B, Human Geography*, 92(1), 1–22.

Boltz, M.G. 2010. Rate and duration memory of naturalistic sounds. *Acta psychologica,* 135(2), 168–81.

Bradbeer, J. 1996. Problem-based learning and fieldwork: A better method of preparation? *Journal of Geography in Higher Education,* 20(1), 11–18.

Brakemeier, M. September 17, 2010 *So klingt Südniedersachsen. "Das authentische Lebensgefühl".* [Online]. Available at: http://www.goettinger-tageblatt.de/Nachrichten/Goettingen/Uebersicht/Das-authentische-Lebensgefuehl [accessed: 17 September 2010].

Brazeal, B. 2003. The music of the Bahian Caboclos. *Anthropological Quarterly*, 76(4), 639–69.

Bridges, B. 2009. *Tan Bodies.* [Audio-guided tour]. Austin.

Bröhl-Kerner, H. 2012. Recorded interview with author, 28 June 2012.

Bruneau, T.J. 1973. Communicative silences: Forms and functions. *Journal of Communication*, 23(1), 17–46.

Bruner, G.C. 1990. Music, mood, and marketing. *The Journal of Marketing*, 54(4), 94–104.

Bruß, S. n.d. *Jede Szene ist ein neuer Anfang. Experiment der Stille. Sound-Design Special.* [Online]. Available at: http://www.experiment-stille.de/gabriel_burns_hoerspiele.php?path=content/inhalt.php&contentid=1356 [accessed: 20 August 2013].

Bull, M. 2000. *Sounding Out the City.* New York: Berg.

Bull, M. 2001a. Soundscapes of the car: A critical study of automobile habitation, in *Car Cultures*, edited by D. Miller. Oxford: Berg, 185–202

Bull, M. 2001b. The world according to sound. *New Media & Society,* 3(2), 179.

Bull, M. 2007. *Sound Moves: iPod Culture and Urban Experience.* New York: Routledge.

Bull, M. and Back, L. 2003. *The Auditory Culture Reader.* New York: Berg.

Bundesministerium der Justiz, 1974. *Gesetz zum Schutz vor schädlichen Umwelteinwirkungen durch Luftverunreinigungen, Geräusche, Erschütterungen und ähnliche Vorgänge.*

Bundesministerium der Justiz, 2006. *Vierunddreißigste Verordnung zur Durchführung des Bundes- Immissionsschutzgesetzes (Verordnung über die Lärmkartierung).*

Burgess, C. and Wathey, A. 2000. Mapping the soundscape: Church music in English towns, 1450–1550. *Early Music History*, 19, 1–46.

Burghardt, P. 2012. Recorded interview with author, 3 July 2012.

Busch, T., Hodgson, M., and Wakefield, C. 2003. Scale-model study of the effectiveness of highway noise barriers. *Journal of the Acoustical Society of America*, 114(4 Pt 1), 1947–54.

Byrne, D.C., Michael, K.L. and Tufts, J.B. 2011. Industrial noise and hearing conservation, in *Patty's Industrial Hygiene*, edited by V.E. Rose and B. Cohrssen. Hoboken, NJ: John Wiley & Sons, 1507–64.

Caldwell, C. and Hibbert, S.A. 2002. The influence of music tempo and musical preference on restaurant patrons' behavior. *Psychology and Marketing*, 19(11), 895–917.

Campaign to Protect Rural England, 2005. *Mapping Tranquility: Defining and Assessing a Valuable Resource.* London: CPRE Publications.

Can, A., Leclercq, L., Lelong, J., and Botteldooren, D. 2010. Traffic noise spectrum analysis: Dynamic modeling vs. experimental observations. *Applied Acoustics*, 71(8), 764–70.

Can, A., Leclercq, L., Lelong, J., and Defrance, J. 2008. Capturing urban traffic noise dynamics through relevant descriptors. *Applied Acoustics*, 69(12), 1270–80.

Carles, J.L., Barrio, I.L., and de Lucio, J.V. 1999. Sound influence on landscape values. *Landscape and Urban Planning*, 43(4), 191–200.

Casey, E.S. 2009. *World, Indiana University Press, Bloomington.* Bloomington and Indianapolis, IN: Indiana University Press.

Cavanaugh, W.J., Tocci, G.C. and Wilkes, J.A. 2010. *Architectural Acoustics: Principles and Practice.* Hoboken, NJ: John Wiley & Sons.

Che, D. 2009. Techno: Music and entrepreneurship in post-Fordist Detroit, in *Sound, Society, and the Geography of Popular Music*, edited by T.L. Bell and O. Johansson. Farnham and Burlington, VT: Ashgate, 261–80.

Chebat, J.C., Chebat, C.G., and Vaillant, D. 2001. Environmental background music and in-store selling. *Journal of Business Research*, 54(2), 115–23.

Christie, M.E. 2004. Kitchenspace, fiestas, and cultural reproduction in Mexican house-lot gardens. *Geographical Review*, 94(3), 368–90.

Cid, C. 2011. Recorded interview with author, 6 October 2011.

City of Austin, 2008. *The Code of the City of Austin, Texas.* Cincinnati, OH: American Legal Publishing Corporation.

Claas, U. 2012. Recorded interview with author, 5 July 2012.

Clark, W.W. and Cox, J.R. 2012. Noise and ultrasound, in *Patty's Toxicology*, edited by E. Bingham and B. Cohrssen. Hoboken, NJ: John Wiley & Sons, 79–108.

Coates, P.A. 2005. The strange stillness of the past: Toward an environmental history of sound and noise. *Environmental History*, 10(4), 636–65.

De Coensel, B., De Muer, T., Yperman, I. and Botteldooren, D. 2005. The influence of traffic flow dynamics on urban soundscapes. *Applied Acoustics*, 66(2), 175–94.

Coensel, B.D., Vanwetswinkel, S., and Botteldooren, D. 2011. Effects of natural sounds on the perception of road traffic noise. *Journal of the Acoustical Society of America,* 129(4), EL148–EL153.

Cohen, E. 1979. A phenomenology of tourist experiences. *Sociology*, 13(2), 179–201.

Committee on Armed Services, 2008. *Inquiry into the Treatment of Detainees in US Custody,* 110th Congress, 2nd Session.

Connell, J. and Gibson, C. 2003. *Sound Tracks: Popular Music, Identity, and Place.* London and New York: Routledge.

Connell, J. and Gibson, C. 2009. Ambient Australia: Music, meditation, and tourist places, in *Sound, Society, and the Geography of Popular Music*, edited by T.L. Bell and O. Johansson. Farnham and Burlington, VT: Ashgate, 67–88.

Council of Europe, Parliamentary Assembly, 2010. *Prohibiting the Marketing and Use of the "Mosquito" Youth Dispersal Device*, Recommendation 1930 (2010).

Cresswell, T. 1996. *In Place/Out of Place: Geography, Ideology, and Transgression.* Minneapolis, MN: University of Minnesota Press.

Cresswell, T. 2004. *Place: A Short Introduction.* Oxford: Wiley-Blackwell.

Curti, G.H. and Craine, J. 2011. Lark's Tongue in Aspect. *Aether*, 7, 137–40.

Dahles, H. 2002. The politics of tour guiding: Image management in Indonesia. *Annals of Tourism Research*, 29(5), 783–800.

Dawson, K.J. 1988. Flight, fancy, and the garden's song. *Landscape Journal*, 7(2), 170–75.

Daynes, S. 2009. A lesson of geography, on the Riddim: The symbolic topography of reggae music, in *Sound, Society, and the Geography of Popular Music*, edited by T.L. Bell and O. Johansson. Farnham and Burlington, VT: Ashgate, 91–105.

Denney, D. 2011. Recorded interview with author, 7 October 2011.

DeNora, T. and Belcher, S. 2000. When you're trying something on you picture yourself in a place where they are playing this kind of music – musically sponsored agency in the British clothing retail sector. *The Sociological Review*, 48(1), 80–101.

Deubel, P. 2007. Podcasts: Improving quality and accessibility. *THE Journal.* [Online]. Available at: http://thejournal.com/articles/2007/06/14/podcasts-improving-quality-and-accessibility.aspx [accessed: 24 July 2013].

Díaz, C. and Pedrero, A. 2006. Sound exposure during daily activities. *Applied Acoustics*, 67(3), 271–83.

Döring, J. and Thielmann, T. 2009. *Mediengeographie.* Bielefeld: transcript Verlag.

Duffy, M. 2000. Lines of drift: Festival participation and performing a sense of place. *Popular Music*, 19(01), 51–64.

Duffy, M. and Waitt, G. 2011. Sound diaries: A method for listening to place. *Aether*, 7, 119–36.

Du Gay, P. 1997. *Doing Cultural Studies: The Story of the Sony Walkman.* London: Sage Publications Ltd.

Eckert, W. 2012. Recorded interview with author, 16 July 2012.

Eco, U. 1998. Die Erzählstruktur bei Ian Fleming, in *Der Kriminalroman. Poetik Theorie Geschichte*, edited by J. Vogt. Munich: W. Fink, 181–207.

Edge, P. 2011. Recorded interview with author, 21 December 2011.

Edler, D. and Lammert-Siepmann, N. 2010. Approaching the acoustic dimension in cartographic theory and practice: An example of mapping Estonia. *Journal for Theoretical Cartography*, 3, 1–15.

Entrikin, J.N. 1991. *The Betweenness of Place: Towards a Geography of Modernity.* London: Macmillan.

Escher, A. 2006. The geography of cinema: A cinematic world. *Erdkunde*, 60(4), 307–14.

Europäisches Parlament und der Rat der Europäischen Union 2002. *ABl. Nr. L 189/12. Richtlinie 2002/49/EG des Europäischen Parlaments und des Rates vom 25. Juni 2002 über die Bewertung und Bekämpfung von Umgebungslärm.*

European Commission, 2004. Research for a Quieter Europe in 2020. *An Updated Startegy Paper of the CALM Network: European Commission, Research Directorate-General*. [Online]. Available at: http://rp7.ffg.at/Kontext/WebService/SecureFileAccess.aspx?fileguid=%7Ba17d7806-1585-43ad-a504-bcbfdcb0da3e%7D [accessed: 27 June 2013].

European Commission Working Group Assessment of Exposure to Noise (WG-AEN) 2006. *Good Practice Guide for Strategic Noise Mapping and the Production of Associated Data on Noise Exposure*. Brüssel: WG-AEN.

Farcas, F. and Sivertunb, 2009. Road Traffic Noise: GIS Tools for Noise Mapping and a Case Study for Skåne Region. *The International Archives of the Photogrammetry, Remote Sensing and Spatial Information Sciences*, 34(XXX), o.S.

Fariss, B. 2009. *Dominion: A Walking Tour.* [Audio-guided tour]. Austin.

Fastl, H. 2004. *Audio-visual Interactions in Loudness Evaluation*. Proceedings of the 18th International Congress on Acoustics, Kyoto, Japan, April, 2004.

Feld, S. 1984. Sound structure as social structure. *Ethnomusicology*, 28(3), 383–409.

Feld, S. and Brenneis, D. 2004. Doing anthropology in sound. *American Ethnologist*, 31(4), 461–74.

Fennell, C.T., Byers-Heinlein, K., and Werker, J.F. 2007. Using speech sounds to guide word learning: The case of bilingual infants. *Child Dev*, 78(5), 1510–25.

Fidell, S., Mestre, V., Schomer, P., Berry, B., Gjestland, T., Vallet, M., and Reid, T. 2011. A first-principles model for estimating the prevalence of annoyance with aircraft noise exposure. *Journal of the Acoustical Society of America*, 130(2), 791–806.

Figgis, M. 2003. Silence: The absence of sound, in *Soundscape: The School of Sound Lectures, 1998–2001*, edited by L. Sider, D. Freeman and J. Sider. London: Wallflower Press, 1–14.

Filimowicz, M. and Stockholm, J. 2010. Towards a phenomenology of the acoustic image. *Organised Sound*, 15(1), 5–12.

Finn, J. and Lukinbeal, C. 2009. Musical Cartographies: Los Ritmos de los Barrios de la Habana, in *Sound, Society, and the Geography of Popular Music*, edited by T.L. Bell and O. Johansson. Farnham and Burlington, VT: Ashgate, 127–44.

Finney, G.L. 1947. Ecstasy and music in seventeenth-century England. *Journal of the History of Ideas*, 8(2), 153–86.

Flitner, M. 2007. *Lärm an der Grenze. Fluglärm und Umweltgerechtigkeit am Beispiel des binationalen Flughafens Basel-Mulhouse.* Stuttgart: Franz Steiner Verlag.

Fontana, B. 1987. The relocation of ambient sound: Urban sound sculpture. *Leonardo*, 20(2), 143–7.

Forsyth, A.J., Barnard, M., and McKeganey, N.P. 1997. Musical preference as an indicator of adolescent drug use. *Addiction*, 92(10), 1317–25.

Fortuna, C. 2001. Soundscapes: The sounding city and urban social life. *Oficina do CES*, 161, 1–16.

Fraedrich, J.P. and King, M.F. 1998. Marketing implications of nonmusical sounds. *Journal of Business and Psychology*, 13(1), 127–39.
Fujikawa, T. and Kobayashi, Y. 2010. The effects of background music and sound in economic decision making: Evidence from a laboratory experiment. *Munich Personal RePEc Archive*, 23374, 1–36.
Gaver, W.W. 1993a. How do we hear in the world? Explorations in ecological acoustics. *Ecological Psychology*, 5(4), 285–313.
Gaver, W.W. 1993b. What in the world do we hear? An ecological approach to auditory source perception. *Ecological Psychology*, 5(1), 1–29.
Ge, J. and Hokao, K. 2005. Applying the methods of image evaluation and spatial analysis to study the sound environment of urban street areas. *Journal of Environmental Psychology*, 25(4), 455–66.
Gere, A.R. 2001. Revealing silence: Rethinking personal writing. *College Composition and Communication*, 53(2), 203–23.
Gerry, D.W., Faux, A.L., and Trainor, L.J. 2010. Effects of Kindermusik training on infants' rhythmic enculturation. *Developmental Science*, 13(3), 545–51.
Gibson, J.J. 1986. *The Ecological Approach to Visual Perception*. London and Hillsdale, NJ: Lawrence Erlbaum.
Gifford, R. and Ng, C.F. 1982. The relative contribution of visual and auditory cues to environmental perception. *Journal of Environmental Psychology*, 2(4), 275–84.
Göldner, H.P. n.d. *Die Deutschen Hörspiel-Charts 2010*. [Online]. Available at: http://www.hoerspiel-box.de/hoerspiel-charts.htm [accessed: 20 August 2013].
Graham, B. and Murray, M. 1997. The spiritual and the profane: The pilgrimage to Santiago de Compostela. *Ecumene*, 4(4), 389–409.
Graves, S. 2009. Hip hop: A postmodem folk music, in *Sound, Society, and the Geography of Popular Music*, edited by T.L. Bell and O. Johansson. Farnham and Burlington, VT: Ashgate, 245–60.
Greated, M. 2011. The nature of sound and vision in relation to colour. *Optics & Laser Technology*, 43(2), 337–47.
Greater London Authority, 2004. *Sounder City: The Mayor's Ambient Noise Strategy*. London: Greater London Authority.
Gregory, D. 1995. Between the book and the lamp: Imaginative geographies of Egypt, 1849–50. *Transactions of the Institute of British Geographers*, 20(1), 29–57.
Guéguen, N., Jacob, C., Lourel, M., and Le Guellec, H. 2007. Effect of background music on consumer's behavior: A field experiment in a open-air market. *European Journal of Scientific Research*, 16(2), 268–72.
Gunderlach, J. 2007. Sound: Exploring a character-defining feature of historic places. *APT Bulletin*, 38(4), 13–20.
Haaf, N. 2003. *Musik als Mittel der Dramaturgie. Hörspielmusik für Kinder und Jugendliche Ende der 70er, Anfang der 80er Jahre*. Diploma thesis [unpublished]. Hamburg.

Hahn, D. 2002. Creating the soundscape for Zagreb everywhere. *Organised Sound*, 7(1), 57–63.

Hahn, T., Pethes, N., and Stäheli, U. 2003. Introduction: Popular noise in global systems. *Soziale Systeme*, 9(2), 205–9.

Haigh, M. and Gold, J.R. 1993. The problems with fieldwork: A group-based approach towards integrating fieldwork into the undergraduate geography curriculum. *Journal of Geography in Higher Education*, 17(1), 21–32.

Halpenny, E.A. 2010. Pro-environmental behaviours and park visitors: The effect of place attachment. *Journal of Environmental Psychology*, 30(4), 409–21.

Harris, E.T. 2005. Silence as sound: Handel's sublime pauses. *Journal of Musicology*, 22(4), 521–58.

Hartmann, H.-J. 2012. Recorded interview with author, 14 July 2012.

Harvey, P. 2009. Grazing: Sense of place. *The Australian Journal of Rural Health*, 17, 286.

Hasse, J. 2012. *Atmosphären der Stadt: aufgespürte Räume.* Berlin: Jovis.

Hausladen, G.J. 1995. Murder in Moscow. *Geographical Review*, 85(1), 63–78.

Health Alliance for Austin Musicians, n.d. *About HAAM.* [Online]. Available at: www.myhaam.org/about/ [accessed: 23 May 2013].

Hedfors, P. 2008. *Site Soundscapes: Landscape Architecture in the Light of Sound.* Doctoral thesis [unpublished]. Uppsala.

Heenan, D.E. 2011. Recorded interview with author, 21 December 2011.

Heiligenthal, S. 2012. Recorded interview with author, 20 June 2012.

Heißenbüttel, H. 1966. *Über Literatur.* Olten und Freiburg im Breisgau: Walter-Verlag.

Hernandez, R.L. 2004. Sacred sound and sacred substance: Church bells and the auditory culture of Russian villages during the Bolshevik Velikii Perelom. *The American Historical Review*, 109(5), 1475–504.

Hetherington, J., Daniel, T.C., and Brown, T.C. 1993. Is motion more important than it sounds?: The medium of presentation in environment perception research. *Journal of Environmental Psychology*, 13(4), 283–91.

Hobson, J., Chisholm, E., and El Refaie, A. 2010. Sound therapy (masking) in the management of tinnitus in adults. *Cochrane Database of Systematic Reviews*, 12, 1–24.

Holmes, R. 2012. *The Mosquito Device and the Human Rights of Children.* Vancouver: British Columbia Civil Liberties Association.

Hosokawa, S. 1984. The walkman effect. *Popular Music*, 4(1), 165–82.

Höfling, K. 2012. Recorded interview with author, 16 August 2012.

Huang, S., Hsu, C.H., and Chan, A. 2010. Tour guide performance and tourist satisfaction: A study of the package tours in Shanghai. *Journal of Hospitality & Tourism Research*, 34(1), 3–33.

Husserl, E. 1960. *Cartesian Meditations: An Introduction to Phenomenology, Translated by Dorion Cairns.* The Hague: Martinus Nijhoff.

Husserl, E. 1973. *Zur Phänomenologie der Intersubjektivität. Texte aus dem Nachlass. Dritter Teil: 1929–1935.* Den Haag: Martinus Nijhoff.

Husserl, E. 1983. *Ideas Pertaining to a Pure Phenomenology and to a Phenomenological Philosophy. First Book. General Introduction to a Pure Phenomenology.* The Hague: Martinus Nijhoff.

Husserl, E. 1990. *Ideas Pertaining to a Pure Phenomenology and to a Phenomenological Philosophy: Second Book, Studies in the Phenomenology of Constitution.* The Hague: Kluwer Academic Publishing.

Husserl, E. 1991. *Ding und Raum: Vorlesungen 1907.* Hamburg: Meiner Verlag.

Husserl, E. 1992a. *Erste Philosophie.* Hamburg: Meiner Verlag.

Husserl, E. 1992b. *Ideen zu einer reinen Phänomenologie und phänomenologischen Philosophie. Erstes Buch. Allgemeine Einführung in die Phänomenologie.* Hamburg: Meiner Verlag.

Husserl, E. 1992c. *Logische Untersuchungen. Zweiter Band. II. Teil. Untersuchungen zur Phänomenologie und Theorie der Erkenntnis.* Hamburg: Meiner Verlag.

Husserl, E. 1995. *Cartesianische Meditationen.* Hamburg: Meiner Verlag.

Husserl, E. 1997. *Experience and Judgement.* Evanston, IL: Northwestern University Press.

Husserl, E. 1999. *Erfahrung und Urteil. Untersuchungen zur Genealogie der Logik.* Hamburg: Meiner Verlag.

Husserl, E. 2001. *Logical investigations. Volume 2.* New York, NY: Routledge.

Husserl, E. 2010. *Thing and space: Lectures of 1907.* Dordrecht: Kluwer.

Iges, J. 2000. Soundscapes: A historical approach. *eContact!*, 3(4). [Online]. Available at: http://cec.concordia.ca/econtact/Histories/HistoricalApproach.htm [accessed: 23 April 2010].

Ihde, D. 2007. *Listening and Voice: Phenomenologies of Sound.* Albany, NY: State University of New York Press.

Imaah, N.O. 2004. Music: A source of inspiration and harmony in architecture – An African view. *International Review of the Aesthetics and Sociology of Music*, 35(2), 169–82.

Ingerslev, F. 1972. Noise and the sound insulation of buildings. *Philosophical Transactions of the Royal Society of London. Series A, Mathematical and Physical Sciences*, 272(1229), 595–602.

Institute for Environment and Health, 1997. *IEH report on the Non-Auditory Effects of Noise.* Norwich: Institute for Environment and Health.

International Committee of the Red Cross, 2007. *ICRC Report on the Treatment of Fourteen "High Value Detainees" in CIA Custody.* Washington, DC: ICRC.

Irwin, A., Hall, D.A., Peters, A., and Plack, C.J. 2011. Listening to urban soundscapes: Physiological validity of perceptual dimensions. *Psychophysiology*, 48(2), 258–68.

Jackiewicz, E. and Craine, J. 2009. Scales of resistance: Billy Bragg and the creation of activist spaces, in *Sound, Society, and the Geography of Popular Music*, edited by T.L. Bell and O. Johansson. Farnham and Burlington, VT: Ashgate, 33–48.

Jackson, A. 1968. Sound and ritual. *Man*, 3(2), 293–99.

Jankowsky, R.C. 2007. Music, spirit possession and the in-between: Ethnomusicological inquiry and the challenge of trance. *Ethnomusicology Forum*, 16(2), 185–208.

Jansen, C. 2012. Recorded Interview with Author, 8 August 2012.

Jazeel, T. 2005. The world is sound? Geography, musicology and British-Asian soundscapes. *Area*, 37(3), 233–41.

Johansson, O. and Bell, T.L. 2009a. *Sound, Society, and the Geography of Popular Music.* Farnham and Burlington, VT: Ashgate.

Johansson, O. and Bell, T.L. 2009b. Where are the new US music scenes?, in *Sound, Society, and the Geography of Popular Music*, edited by O. Johansson and T.L. Bell. Farnham and Burlington, VT: Ashgate, 219–39.

Johnson, B. and Cloonan, M. 2009. *Dark Side of the Tune: Popular Music and Violence.* Burlington, VT: Ashgate.

Johnson, T. 2011. Salsa politics. *Aether*, 7, 97–118.

Jones, D.M., Chapman, A.J., and Auburn, T.C. 1981. Noise in the environment: A social perspective. *Journal of Environmental Psychology*, (1), 43–59.

Jones, S. 2006. Space-dis-place: How sound and interactivity can reconfigure our apprehension of space. *Leonardo Music Journal*, 16, 20–27.

Jorgensen, B.S. and Stedman, R.C. 2001. Sense of place as an attitude: Lakeshore owners attitudes toward their properties. *Journal of Environmental Psychology*, 21(3), 233–48.

Kang, J. 2007. *Urban Sound Environment.* New York, NY: Spon Press.

Kania, A. 2010. Silent music. *Journal of Aesthetics and Art Criticism*, 68(4), 343–53.

Kant, E. 2007. *Critique of pure reason.* New York, NY: Dutton.

Kastner, S. 2008. *Klang macht Marken.* Wiesbaden: Dr. Th. Gabler.

Keller, A. 2009. *Curator's Notes.* [Audio-guided tour]. Austin.

Kelner, S. and Sanders, G. 2009. Beyond the field trip teaching tourism through tours. *Teaching Sociology*, 37(2), 136–50.

Kennedy, C.B., Sell, J.L., and Zube, E.H. 1988. Landscape aesthetics and geography. *Environmental Review: ER,* 12(3), 31–55.

Keough, S.B. 2009. Internet Radio and Cultural Connections: A Case Study of the St. John's, Newfoundland Radio Market, in *Sound, society, and the geography of popular music*, edited by T.L. Bell and O. Johansson. Farnham and Burlington, VT: Ashgate, 185–203.

Keough, S.B. 2011. Promoting and preserving cultural identity through Newfoundland radio music broadcasts. *Aether,* 7, 75–96.

Kianicka, S., Buchecker, M., Hunziker, M., and Müller-Böker, U. 2006a. Locals' and Tourists' Sense of Place. *Mountain Research and Development,* 26(1), 55–63.

Kianicka, S., Buchecker, M., Hunziker, M., and Müller-Böker, U. 2006b. Locals' and Tourists' Sense of Place. *Mountain Research and Development,* 26(1), 55–63.

Kim, K.-H. and Iwamiya, S.-I. 2008. Formal Congruency between Telop Patterns and Sound Effects. *Music Perception,* 25(5), 429–448.

Kim, P. and Rowland, N.J. n.d. *Sounds like Sociology: A Treatise on the Social Implications of Sound.* Paper [unpublished]. Bloomington.

Kleilein, D. 2008. *Tuned city: zwischen Klang- und Raumspekulation.* Idstein: Kookbooks.

Kluizenaar, Y., Salomons, E.M., Janssen, S.A., van Lenthe, F.J., Vos, H., Zhou, H., Miedema, H.M., and Mackenbach, J.P. 2011. Urban road traffic noise and annoyance: The effect of a quiet façade. *Journal of the Acoustical Society of America*, 130(4), 1936–1942.

Kobayashi, M. and Schmandt, C. 1997. Dynamic Soundscape: mapping time to space for audio browsing. *CHI*, (97), 194–201.

Kornfeld, A.L. 2008. Die kartographische Visualisierung des akustischen Raums. *Kartographische Nachrichten*, 58(6), 294–301.

Kornfeld, A.L., Schiewe, J., and Dykes, J. 2011. Audio Cartography: Visual Encoding of Acoustic Parameters, in *Advances in Cartography and GIScience. Volume 1. Selection from ICC 2011, Paris*, edited by A. Ruas. Berlin and Heidelberg: Springer, 13–31.

Köchy, S. 2005. *The Three Investigators auf Französisch und Deutsch.* Diploma thesis [unpublished]. Germersheim.

Kracauer, S. 1979. *Der Detektiv-Roman : Ein philosophischer Traktat.* Frankfurt am Main: Suhrkamp.

Kristjánsdóttir, Ó. and Kristjánsdóttir, G. 2011. Randomized clinical trial of musical distraction with and without headphones for adolescents' immunization pain. *Scandinavian Journal of Caring Sciences,* 25(1), 19–26.

Kruse, H. 2009. Local Independent Music Scenes and the Implications of the Internet, in *Sound, society, and the geography of popular music,* edited by T.L. Bell and O. Johansson. Farnham and Burlington, VT: Ashgate, 205–217.

Kruse II, R.J. 2009. Geographies of John and Yoko's 1969 Campaign for Peace: An Intersection of Celebrity, Space, Art, and Activism, in *Sound, society, and the geography of popular music*, edited by T.L. Bell and O. Johansson. Farnham and Burlington, VT: Ashgate, 11–31.

Krygier, J.B. 1994. Sound and geographic visualization, in *Visualization in modern cartography,* edited by A. MacEachren and D.R.F. Taylor. New York: Pergamon, 149–166.

Kryter, K.D. 1972. Non-auditory effects of environmental noise. *American Journal of Public Health,* 62(3), 389–398.

Kryter, K.D. 2007. Acoustical, sensory, and psychological research data and procedures for their use in predicting effects of environmental noises. *Journal of the Acoustical Society of America*, 122(5), 2601–2614.

Kuhlke, O. 2009. The Geography of "Canadian Shield Rock": Locality, Nationality and Place Imagery in the Music of the Rheostatics, in *Sound, Society, and the Geography of Popular Music*, edited by T.L. Bell and O. Johansson. Farnham and Burlington, VT: Ashgate, 161–179.

Kun, J.D. 2000. The Aural Border. *Theatre Journal,* 52(1), 1–21.

Kurakula, V. 2007. *A GIS-Based Approach for 3D Noise Modelling Using 3D City Models.* Master's thesis [unpublished]. Enschede.

Kurra, S. and Dal, L. 2011. Sound insulation design by using noise maps. *Building and Environment,* in press.

Kyle, G. and Chick, G. 2007. The social construction of a sense of place. *Leisure Sciences*, 29(3), 209–225.

Lady Bird Johnson Wildflower Center 2009. *Audio Tour of the Lady Bird Johnson Wildflower Center.* [Audio-guided tour]. Austin.

Landmarks n.d. *Audio Tour for Works of Public Art.* [Audio-guided tour]. Austin.

Lanza, J. 1991. The sound of cottage cheese (Why background music is the real world beat!). *Performing arts journal,* 13(3), 42–53.

Lavandier, C., Barbot, B., Terroir, J., and Schuette, M. 2011. Impact of aircraft sound quality combined with the repetition of aircraft flyovers on annoyance as perceived activity disturbance in a laboratory context. *Applied Acoustics,* 72(4), 169–176.

Law, C., Lee, C., Lui, A.S., Yeung, M.K., and Lam, K. 2011. Advancement of three-dimensional noise mapping in Hong Kong. *Applied Acoustics,* 72(8), 534–543.

Leasure, J. 2011. Recorded Interview with Author, 16 December 2011.

Lebiedowska, B. 2005. Acoustic background and transport noise in urbanised areas: A note on the relative classification of the city soundscape. *Transportation Research Part D: Transport and Environment,* 10(4), 341–345.

Leyshon, A., Matless, D., and Revill, G. 1995. The Place of Music: [Introduction]. *Transactions of the Institute of British Geographers,* 120(4), 423–433.

Lim, C., Kim, J., Hong, J., and Lee, S. 2008. Effect of background noise levels on community annoyance from aircraft noise. *Journal of the Acoustical Society of America,* 123(2), 766–771.

Lindenbaum, J. 2009. The Production of Contemporary Christian Music: A Geographical Perspective, in *Sound, society, and the geography of popular music,* edited by T.L. Bell and O. Johansson. Farnham and Burlington, VT: Ashgate, 281–294.

Locke, J. 2000. *An essay concerning human understanding.* London: J M Dent.

Long, J. 2010. *Weird city: Sense of place and creative resistance in Austin, Texas.* Austin: University of Texas Press.

Lukinbeal, C. and Zimmermann, S. 2006. Film Geography: A New Subfield. *Erdkunde,* 60(4), 315–325.

Lukinbeal, C. and Zimmermann, S. 2008. *The Geography of Cinema - A Cinematic World.* Stuttgart: Steiner.

Macdonald, S. 2006. Mediating heritage Tour guides at the former Nazi Party Rally Grounds, Nuremberg. *Tourist studies,* 6(2), 119–138.

Makarewicz, R. and Galuszka, M. 2011. Road traffic noise prediction based on speed-flow diagram. *Applied Acoustics,* 72(4), 190–195.

Makarewicz, R. and Zółtowski, M. 2008. Variations of road traffic noise in residential areas. *Journal of the Acoustical Society of America,* 124(6), 3568–3575.

Maris, E., Stallen, P.J., Vermunt, R., and Steensma, H. 2007. Evaluating noise in social context: the effect of procedural unfairness on noise annoyance judgments. *Journal of the Acoustical Society of America,* 122(6), 3483–3494.

Martinez, M. 2011. No cash necessary for *Weird Science*, aerial dancing and more. *Austin American Statesman,* 14 November, D2.

Massey, D. 1991. A global sense of place. *Marxism today,* 35(6), 24–29.

Massey, D. 1994. *Space, place, and gender.* Cambridge: Polity Press.

Massey, D., Allen, J. and Pile, S. 2000. *City worlds.* London and New York: Routledge.

Matless, D. 2005. Sonic geography in a nature region. *Social & Cultural Geography,* 6(5), 745–766.

Maturana, H.R. 2001. *Was ist erkennen? : die Welt entsteht im Auge des Betrachters.* Munich: Goldmann.

Maturana, H.R. and Varela, F.J. 1992. *The tree of knowledge: the biological roots of human understanding.* Boston: Shambhala.

McCartney, A. 2000. Soundscape Composition and the Subversion of Electroacoustic Norms. *eContact!* [Online], 3(4), Available at: http://cec.concordia.ca/econtact/Histories/SoundscapeComposition.htm [accessed: 16 April 2010].

McDonald, M. 2011. ACL not accessible to all. *Austin American Statesman,* 25 September, D6.

McDonell, G. 2011. Recorded Interview with Author, 8 November 2011.

Meier-Dallach, H.P. 2007. Die Töne der Globalisierung. Beobachtungen aus Zürich. *dérive – Zeitschrift für Stadtforschung* [Online], 27, Available at: http://www.derive.at/index.php?p_case=2&id_cont=608&issue_No=27 [accessed: 16 April 2010].

Mellard, A. 2011. Recorded Interview with Author, 5 October 2011.

Metzer, D. 2006. Modern Silence. *Journal of Musicology,* 23(3), 331–374.

Meyrowitz, J. 1985. *No sense of place: The impact of electronic media on social behavior.* New York: Oxford University Press.

Miedema, H.M. and Vos, H. 2003. Noise sensitivity and reactions to noise and other environmental conditions. *Journal of the Acoustical Society of America,* 113(3), 1492–1504.

Miller, D.W. and Marks, L.J. 1992. Mental imagery and sound effects in radio commercials. *Journal of Advertising,* 21(4), 83–93.

Milliman, R.E. 1982. Using background music to affect the behavior of supermarket shoppers. *The journal of Marketing,* 46(3), 86–91.

Milliman, R.E. 1986. The influence of background music on the behavior of restaurant patrons. *Journal of Consumer Research,* 13(2), 286–289.

Miracle Recreation Equipment Company n.d. *Miracle Tech Security Products.* [Online]. Available at: http://www.miracle-recreation.com/products/miracle-tech-security-products.html [accessed: 26 February 2013].

Mitchell Pierce, K. and Short, K.G. 1994. Children's Books. A sense of place. *The Reading Teacher,* 47(6), 490–497.

Mongillo, P. 2011a. Celebratory memorial to blues musician Bramhall. *Austin American Statesman,* 15 November, B1.

Mongillo, P. 2011b. Dust, drama don‘t put damper on festival. *Austin American Statesman,* 6 November, D1.

Mongillo, P. 2011c. Musicians to hold concert for victims who ‘don’t have a lot’. *Austin American Statesman*, 22 September, B1.

Mongillo, P. 2011d. Sweet November. *Austin American Statesman,* 1 November, D1.

Morley, V. and Somdahl-Sands, K. 2011. Music with a Message. *Aether,* 7, 58–74.

Murphy, T.P. 1996. Sense-making. A case study using the movie Jurassic Park. *Journal of Science Education and Technology,* 5(1), 77–86.

Murray, D. 2011. Recorded Interview with Author, 22 December 2011.

Müller, S. 2012. Recorded Interview with Author, 28 July 2012.

Mylott, E. and Dubois Williams, K. 2007. Quiet Places. *Metroscape,* (Summer), 4–5.

Nelson, L.P. 2007. Sensing the Sacred. *Winterthur Portfolio,* 41(4), 203–238.

Neumann, H. n.d. *Die Drei Fragezeichen. Phonophobia – Synphonie der Angst. Programminformation.* [Online]. Available at: http://www.hn-pr.de/diedrei_files/PM%20DDF%202014_04.pdf [accessed: 20 August 2013].

Neustadt, R. 2004. Music as memory and torture: Sounds of repression and protest in Chile and Argentina. *Chasqui,* 33(1), 128–137.

Newman, R. 1960. Acoustics. *Journal of Architectural Education,* 16(2), 26–33.

Ng, C.F. 2000. Effects of building construction noise on residents: A quasi-experiment. *Journal of Environmental Psychology*, 20(4), 375–385.

Ng, C.F. 2003. Satisfying shoppers’ psychological needs: From public market to cyber-mall* 1. *Journal of Environmental Psychology*, 23(4), 439–455.

Nilo, E.R. 1969. Needs of the Hearing Impaired. *The American Journal of Nursing*, 69(1), 114–116.

O´Meara, C.P. and Tretter, E.M. 2013. Sounding Austin. Live Music, Race, and the Selling of a City, in *Musical Performance and the Changing City. Post-industrial Contexts in Europe and the United States,* edited by F. Holt and C. Wergin. London: Routledge, 52–76.

Öhrström, E., Hadzibajramovic, E., Holmes, M., and Svensson, H. 2006. Effects of road traffic noise on sleep: Studies on children and adults. *Journal of environmental psychology,* 26(2), 116–126.

Pamanikabud, P. 2009. Impact Assessment of Motorway Traffic Noise Using Visualized Noise Mapping Technique. *Crete, WSEAS Transactions on Environment and Development*, 1(5), 24–33.

Pamanikabud, P. and Tansatcha, M. 2010. 3D analysis and investigation of traffic noise impact from a new motorway on building and surrounding area. *Applied Acoustics*, 71(12), 1185–1193.

Papadimitriou, K., Mazaris, A., Kallimanis, A., and Pantis, J. 2009. Cartographic representation of the sonic environment. *Cartographic Journal, The*, 46(2), 126–135.

Pedersen, E. and Larsman, P. 2008. The impact of visual factors on noise annoyance among people living in the vicinity of wind turbines. *Journal of Environmental Psychology,* 28(4), 379–389.

Pesses, M.W. 2009. The City She Loves Me: The Los Angeles of the Red Hot Chili Peppers, in *Sound, society, and the geography of popular music*, edited by T.L. Bell and O. Johansson. Farnham and Burlington, VT: Ashgate, 145–159.

Pesses, M.W. 2011. What's He Building in There? *Aether*, 7, 42–57.

Peterson, I. 1996. Sounds of the seasons. *Science News*, 150(25/26), 400–401.

Petkus, C. 2009. *Recreating the Domain.* [Audio-guided tour]. Austin.

Pheasant, R.J., Fisher, M.N., Watts, G.R., Whitaker, D.J., and Horoshenkov, K.V. 2010. The importance of auditory-visual interaction in the construction of 'tranquil space'. *Journal of Environmental Psychology*, accepted manuscript.

Pheasant, R.J., Horoshenkov, K., Watts, G., and Barrett, B. 2008. The acoustic and visual factors influencing the construction of tranquil space in urban and rural environments tranquil spaces-quiet places? *Journal of the Acoustical Society of America*, 123(3), 1446–1457.

Piaget, J. 1974. *Biology and knowledge.* Chicago: University of Chicago Press.

Piaget, J. 1976. *Die* Äquilibration *der kognitiven Strukturen.* Stuttgart: Ernst Klett.

Piaget, J. 1992. *Einführung in die genetische Erkenntnistheorie.* Frankfurt am Main: Suhrkamp.

Piaget, J. 1997. *The moral judgment of the child.* New York: Free Press.

Piaget, J. 2007. *The child's conception of the world.* Lanham, Maryland: Rowman & Littlefield.

Pizam, A. 1999. The American group tourist as viewed by British, Israeli, Korean, and Dutch tour guides. *Journal of Travel Research*, 38(2), 119–126.

Plato 2008. *Republic.* New York: Oxford University Press.

Pocock, D. 1996. Place evocation: the Galilee chapel in Durham Cathedral. *Transactions of the Institute of British Geographers*, 21(2), 379–386.

Raimbault, M., Lavandier, C., and Bérengier, M. 2003. Ambient sound assessment of urban environments: Field studies in two French cities. *Applied Acoustics*, 64(12), 1241–56.

Raimbault, M. and Dubois, D. 2005. Urban soundscapes: Experiences and knowledge. *Cities*, 22(5), 339–50.

Ramis, J., Alba, J., Garcia, D., and Hernandez, F. 2003. Noise effects of reducing traffic flow through a Spanish city. *Applied Acoustics*, 64(3), 343–64.

Rasmussen, B. 2010. Sound insulation between dwellings-requirements in building regulations in Europe. *Applied Acoustics*, 71(4), 373–85.

Raymond, C.M., Brown, G., and Weber, D. 2010. The measurement of place attachment: Personal, community, and environmental connections. *Journal of Environmental Psychology*, 30(4), 422–34.

Reining, S. 2012. Recorded interview with author, 25 June 2012.

Relph, E.C. 1976. *Place and Placelessness.* London: Pion.

Revzin, I.I. 1998. Zur semiotischenAnalyse des Detektivromans am Beispiel der RomaneAgathaChristies, in *Der Kriminalroman. Poetik Theorie Geschichte*, edited by J. Vogt. Munich: W. Fink, 154–6.

Reyes, R. 2011. Recorded interview with author, 20 December 2011.

Richoux, O., Ayrault, C., Pelat, A., Félix, S., and Lihoreau, B. 2010. Effect of the open roof on low frequency acoustic propagation in street canyons. *Applied Acoustics*, 71(10), 731–8.

Robinson, J. and Hatten, R.S. 2012. Emotions in music. *Music Theory Spectrum*, 34(2), 71–106.

Robinson, P. 2009. *Border's Card.* [Audio-guided tour]. Austin.

Rodaway, P. 1994. *Sensuous Geographies: Body, Sense and Place.* London und New York, NY: Routledge.

Rodermond, J. 2005. *Die drei??? & Co* – über *die Bedeutung von Musik in Kinder- und Jugendhörspielserien.* Seminar paper [unpublished]. Bremen.

Roizen, N.J. 2003. Nongenetic causes of hearing loss. *Mental Retardation and Developmental Disabilities Research Reviews*, 9(2), 120–27.

Rollero, C. and De Piccoli, N. 2010. Place attachment, identification and environment perception: An empirical study. *Journal of Environmental Psychology*, 30(2), 198–205.

Romig, K. 2009. A listener's mental map of California, in *Sound, Society, and the Geography of Popular Music*, edited by T.L. Bell and O. Johansson. Farnham and Burlington, VT: Ashgate, 107–21.

Rose, G. 1995. Place and identity: A sense of place, in *A Place in the World*, edited by D. Massey and J. Pat. New York, NY: Oxford University Press, 87–132.

Rose, G. 2012. *Visual Methodologies: An Introduction to the Interpretation of Visual Materials.* London: Sage Publications Ltd.

Rossetto, V. 2009. *Century Oaks.* [Audio-guided tour]. Austin.

Rouget, G. 1985. *Music and Trance: A Theory of the Relations Between Music and Possession.* Chicago, IL and London: University of Chicago Press.

Rychtáriková, M. and Vermeir, G. 2013. Soundscape categorization on the basis of objective acoustical parameters. *Applied Acoustics*, 74, 240–47.

Ryden, K.C. 1993. *Mapping the Invisible Landscape: Folklore, Writing, and the Sense of Place.* Iowa City, IA: University of Iowa Press.

Said, E.W. 2003. *Orientalism.* London: Penguin.

Sakakeeny, M. 2010. "Under the Bridge": An orientation to soundscapes in New Orleans. *Ethnomusicology*, 54(1), 1–27.

Salomons, E.M. and Berghauser Pont, M. 2012. Urban traffic noise and the relation to urban density, form, and traffic elasticity. *Landscape and Urban Planning*, 108(1), 2–16.

Save our Springs Alliance, 2010. *Barton Springs Audio Tour.* [Audio-guided tour]. Austin.

Schafer, R.M. 1969. *The New Soundscape.* Scarborough, Ontario: Berandol Music Limited.

Schafer, R.M. 1970. *The Book of Noise.* Wellington: Price Milburn and Company Limited.

Schafer, R.M. 1994. *The Soundscape: Our Sonic Environment and the Tuning of the World.* Rochester, VT: Destiny Books.

Schäfer, K. 2006. *Empfehlungen für die Durchführung schalltechnischer Untersuchungen als Teil der wasserbaulichen Planung.* Bonn: Bundesministerium für Verkehr, Bau und Stadtentwicklung.

Schiewe, J. and Kornfeld, A.L. 2009. *Framework and Potential Implementations of Urban Sound Cartography.* Proceedings of the 12th AGILE International Conference on Geographic Information Science, Hannover, June, 2009.

Schlottmann, A. and Miggelbrink, J. 2009. Visual geographies – an editorial. *Social Geography*, 4, 1–11. [Online]. Available at: http://www.soc-geogr.net/4/1/2009/sg-4-1-2009.pdf [accessed: 20 August 2013].

Schmidt, K. 2002. *Die drei ??? - Eine populäre Kriminalserie für Kinder. Untersuchungen zur Konzeption und medienübergreifenden Vermarktung.* Diploma thesis [unpublished]. Stuttgart.

Schnell, R., Hill, P.B. and Esser, E. 2011. *Methoden der empirischen Sozialforschung.* Munich: Oldenbourg Verlag.

Schnell, S.M. 2000. The Kiowa Homeland in Oklahoma. *Geographical Review*, 90(2), 155–76.

Schryer, C. 1998. À la recherche de l›effet Sharawadji/Electroacoustic Soundscape Composition. *eContact!*, 1(4). [Online]. Available at: http://cec.concordia.ca/econtact/Ecology/Schryer.html [accessed: 26 February 2010].

Schulte-Fortkamp, B. and Genuit, K. 2004. The acoustical diary as an innovative tool in soundscape evaluation. *The Journal of the Acoustical Society of America*, 115(5), 2496.

Schyma, B. 2002. *Neue Elektronische Musik in Köln: Szenen, Ströme und Kulturen.* Master's thesis [unpublished]. Köln.

Servigne, S., Laurini, R., Kang, M.A., and Li, K.J. 1999. *First Specifications of an Information System for Urban Soundscape.* Procedings of the IEEE International Conference on Multimedia Computing and Systems, Lyon, January, 1999.

Shamai, S. and Ilatov, Z. 2005. Measuring sense of place: Methodological aspects. *Tijdschrift voor economische en sociale geografie,* 96(5), 467–76.

Simmel, G. 1908. *Soziologie: Untersuchungen* über *die Formen der Vergesellschaftung.* Berlin: Duncker & Humblot.

Simun, M. 2009. My music, my world: Using the MP3 player to shape experience in London. *New Media & Society*, 11(6), 921.

Sklovskij, V. 1998. Die Kriminalerzählung bei Conan Doyle, in *Der Kriminalroman. Poetik Theorie Geschichte*, edited by J. Vogt. Munich: W. Fink, 142–53.

Smith, A.P. 1991. A review of the non-auditory effects of noise on health. *Work & Stress*, 5(1), 49–62.

Smith, A.P. and Broadbent, D.E. 1991. *Non-auditory Effects of Noise at Work: A Review of the Literature.* Bootle: Health and Safety Executive.

Sony Music Entertainment Germany GmbH, n.d. *Europa fact sheet.* [Online]. Available at: http://www.natuerlichvoneuropa.de/media/files/europa-fact-sheetstand-2010_268.pdf [accessed: 20 August 2013].

Stansfeld, S.A., Clark, C., Cameron, R.M., Alfred, T., Head, J., Haines, M.M., van Kamp, I., van Kempen, E., and Lopez-Barrio, I. 2009. Aircraft and road traffic noise exposure and children's mental health. *Journal of Environmental Psychology*, 29(2), 203–7.

Stansfeld, S.A. and Matheson, M.P. 2003. Noise pollution: Non-auditory effects on health. *British Medical Bulletin*, 68(1), 243–57.

Stevens, D. 2011. Recorded interview with author, 22 November 2011.

Szafera, L. 2012. Recorded interview with author, 3 July 2012.

Tagg, P. 1994. Subjectivity and soundscape, motorbikes and music, in *Soundscapes: Essays on Vroom and Moo*, edited by H. Järviluoma. Tampere: Tampere University Press, 48–66.

Tallon, L. n.d. *About that 1952 Sedelijk Museum Audio Guide, and a Certain Willem Sandburg.* [Online]. Available at: http://musematic.net/2009/05/19/about-that-1952-sedelijk-museum-audio-guide-and-a-certain-willem-sandburg/ [accessed: 19 May 2009].

Tassi, P., Rohmer, O., Bonnefond, A., Margiocchi, F., Poisson, F., and Schimchowitsch, S. 2012. Long term exposure to nocturnal railway noise produces chronic signs of cognitive deficits and diurnal sleepiness. *Journal of Environmental Psychology*, 33, 45–52.

Texas Department of Transportation, 2011. *Guidelines for Analysis and Abatement of Roadway Traffic Noise.* Austin, TX: Texas Department of Transportation.

Texas Legislature, 2007. *Penal Code. Title 9. Offenses Against Public Order And Decency. Chapter 42. Disorderly Conduct And Related Offenses.*

Texas Memorial Museum 2011. *Guide By Cell Report, September 30, 2007 to October 6, 2011.* Austin, TX: Texas Memorial Museum.

Texas State Cemetery, n.d. *Texas State Cemetery Tour.* [Audio-guided tour]. Austin.

Texas State Travel Agency, 2012. *Austin Walking Tour.* [Audio-guided tour]. Austin.

The City of Austin, Texas, 1992. *The Code of the City of Austin, Texas. § 9–2 Noise and Amplified Sound.* Austin, TX: American Legal Publishing Corporation.

Thompson, E. 2004. *The Soundscape of Modernity: Architectural Acoustics and the Culture of Listening in America, 1900–1933.* Cambridge, MA and London: The MIT Press.

Thomson, P. 2005. Soundscape composition, globality, and implicated critique. *eContact!*, 8(3). [Online]. Available at: http://cec.concordia.ca/econtact/8_3/thomson.html [accessed: 23 April 2010].

Thurston, F.E. 2013. The worker's ear: A history of noise-induced hearing loss. *American Journal of Industrial Medicine*, 56(3), 367–77.

Todorov, T. 1998. Typologie des kriminalromans, in *Der Kriminalroman: Poetik Theorie Geschichte*, edited by J. Vogt. Munich: W. Fink, 208–15.

Truax, B. 1978. *Handbook for Acoustic Ecology.* Cambridge: ARC Publications, Vancouver, BC.

Truax, B. 1996. Soundscape, acoustic communication and environmental sound composition. *Contemporary Music Review*, 15(1), 49–65.

Truax, B. 2001. *Acoustic Communication.* Westport, CT: Ablex Publishing.

Truax, B. 2002. Genres and techniques of soundscape composition as developed at Simon Fraser University. *Organised Sound*, 7(1), 5–14.

Tsai, K.T., Lin, M.D., and Chen, Y.H. 2009. Noise mapping in urban environments: A Taiwan study. *Applied Acoustics*, 70(7), 964–72.

Tuan, Y.F. 1974. *Topophilia: A Study of Environmental Perception, Attitudes, and Values*. New York, NY: Columbia University Press.

Tuan, Y.F. 1975. Place: An experiential perspective. *Geographical Review*, 65(2), 151–65.

Tuan, Y.F. 1977. *Space and Place: The Perspective of Experience*. Minneapolis, MN: University of Minnesota Press.

Tuan, Y.F. 1986. Strangers and strangeness. *Geographical Review*, 76(1), 10–19.

Tuan, Y.F. 1991. Language and the making of place: A narrative-descriptive approach. *Annals of the Association of American Geographers*, 81(4), 684–96.

Tuan, Y.F. 1994. The city and human speech. *Geographical Review*, 84(2), 144–51.

Tuan, Y.F. 2001. Life as a field trip. *Geographical Review*, 91(1–2), 41–5.

Turley, L.W. and Milliman, R.E. 2000. Atmospheric effects on shopping behavior: a review of the experimental evidence. *Journal of Business Research*, 49(2), 193–211.

TXP, Inc., 2012. *The Economic Impact of the Creative Sector in Austin – 2012 Update*. Austin, TX: TXP, Inc.

US Department of Transportation, 1995. *Highway Traffic Noise Analysis and Abatement Policy and Guidance*. Washington, DC: US Department of Transportation.

Valentine, G. 1989. The geography of women's fear. *Area*, 21(4), 385–90.

Vallejo, A. 2011. Recorded interview with author, 14 September 2011.

Vanel, H. 2008. John Cage's Muzak-Plus: The fu(rni)ture of music. *Representations*, 102(1), 94–128.

van Kempen, E., van Kamp, I., Nilsson, M., Lammers, J., Emmen, H., Clark, C., and Stansfeld, S. 2010. The role of annoyance in the relation between transportation noise and children's health and cognition. *Journal of the Acoustical Society of America*, 128(5), 2817–28.

Van Ryzin, J.C. 2011. The return of the 'Intergalactic Nemesis'. *Austin American Statesman*, 21 November, D2.

Velasco, D. 2000. Island landscape: Following in Humboldt's footsteps through the acoustic spaces of the tropics. *Leonardo Music Journal*, 10, 21–4.

Viollon, S., Lavandier, C., and Drake, C. 2002. Influence of visual setting on sound ratings in an urban environment. *Applied Acoustics*, 63(5), 493–511.

Visell, Y., Fontana, F., Giordano, B.L., Nordahl, R., Serafin, S., and Bresin, R. 2009. Sound design and perception in walking interactions. *International Journal of Human-computer Studies*, 67(11), 947–59.

Vogt, J. 1998. *Der Kriminalroman. Poetik Theorie Geschichte*. Munich: W. Fink.

von Foerster, H. 2000. *Einführung in den Konstruktivismus*. Munich: Piper.

von Glasersfeld, E. 1989. An exposition of constructivism: Why some like it radical. *Constructivist Views on the Teaching and Learning of Mathematics: JRME Monographs*. [Online]. Available at: http://www.oikos.org/constructivism.htm [accessed: 22 August 2012].

von Glasersfeld, E. 1995. *Radical Constructivism: A Way of Knowing and Learning*. London: Routledge.

von Glasersfeld, E. 1997. Distinguishing the observer: An attempt at interpreting Maturana. *Towards an Ecology of Mind*. [Online]. Available at: http://www.oikos.org/vonobserv.htm [accessed: 21 August 2012].

Waldenfels, B. 1994. *Antwortregister*. Frankfurt am Main: Suhrkamp.

Waldenfels, B. 1997. *Topographie des Fremden*. Frankfurt am Main: Suhrkamp.

Waldenfels, B. 1998. *Grenzen der Normalisierung*. Frankfurt am Main: Suhrkamp.

Waldenfels, B. 1999a. *Sinnesschwellen*. Frankfurt am Main: Suhrkamp.

Waldenfels, B. 1999b. *Vielstimmigkeit der Rede*. Frankfurt am Main: Suhrkamp.

Waldenfels, B. 2000. *Das leibliche Selbst: Vorlesungen zur Phänomenologie des Leibes*. Frankfurt am Main: Suhrkamp.

Waldenfels, B. 2002. *Bruchlinien der Erfahrung*. Frankfurt am Main: Suhrkamp.

Waldenfels, B. 2004. *Phänomenologie der Aufmerksamkeit*. Frankfurt am Main: Suhrkamp .

Waldenfels, B. 2005. *In den Netzen der Lebenswelt*. Frankfurt am Main: Suhrkamp.

Waldenfels, B. 2006. *Grundmotive einer Phänomenologie des Fremden*. Frankfurt am Main: Suhrkamp.

Waldenfels, B. 2010. *Sinne und Künste im Wechselspiel: Modi* ästhetischer *Erfahrung*. Frankfurt am Main: Suhrkamp.

Waldenfels, B. 2011. *Phenomenology of the Alien: Basic Concepts*. Evanston, IL: Northwestern University Press.

Walerian, E., Janczur, R., Czechowicz, M., and Smyrnova, Y. 2011. Validation of a complex urban noise model close to a road. *Applied Acoustics*, 72(10), 790–802.

Waterman, E. 2000. Sound escape: Sonic geography remembered and imagined. *Cultural Geographies*, 7(1), 112–15.

Watts, G., Chinn, L., and Godfrey, N. 1999. The effects of vegetation on the perception of traffic noise. *Applied Acoustics*, 56(1), 39–56.

Watzlawick, P. 1995. *Vom Unsinn des Sinns oder vom Sinn des Unsinns*. München und Zürich: Piper.

Watzlawick, P. 1998. *Wie wirklich ist die Wirklichkeit? : Wahn, Täuschung, Verstehen*. Munich: Piper.

Werner, H.U. 2006. *Soundscape-Dialog*. Göttingen: Vandenhoeck & Ruprecht.

West, G. 2011. Recorded interview with author, 5 October 2011.

Westerkamp, H. 1974. Soundwalking. *Sound Heritage*, 3(4), 18–27.

Westerkamp, H. 1988. *Listening and Soundmaking: A Study of Music-as-Environment*. Burnaby: Simon Fraser University.

Wigbers, M. 2006. *Krimi-Orte im Wandel: Gestaltung und Funktionen der Handlungsschauplätze in Kriminalerzählungen von der Romantik bis in die Gegenwart*. Würzburg: Königshausen & Neumann.

Wilson, S. 2003. The effect of music on perceived atmosphere and purchase intentions in a restaurant. *Psychology of Music*, 31(1), 93–112.

Windsor, J.E. and McVey, J.A. 2005. Annihilation of both place and sense of place: The experience of the Cheslatta T'En Canadian First Nation within the context of large-scale environmental projects. *The Geographical Journal*, 171(2), 146–65.

Winkelmann, S. 2012. Recorded interview with author, 25 June 2012.

Wissmann, T. 2010. Field notebook Lisbon, 15 March 2010.

Wissmann, T. 2011. *Raum zur Identitätskonstruktion des Eigenen.* Stuttgart: Franz Steiner Verlag.

Wissmann, T. 2013. Tuning the field trip: Audio-guided tours as a replacement for 1–day excursions in human geography. *Journal of Geography in Higher Education*, 1–16. [Online]. Available at: http://dx.doi.org/10.1080/03098265.2013.828684 [accessed: 30 August 2013].

Wissmann, T. and Zimmermann, S. 2010. From hear to there: Sound and the congnitive construction of world in popular audioplays. *Erdkunde*, 64(4), 371–83.

Wood, R.M. and Gustafson, G.E. 2001. Infant crying and adults' anticipated caregiving responses: Acoustic and contextual influences. *Child Dev*, 72(5), 1287–300.

Wynn, J.R. 2005. Guiding practices: Storytelling tricks for reproducing the urban landscape. *Qualitative Sociology*, 28(4), 399–417.

Wynn, J.R. 2010. City Tour guides: Urban alchemists at work. *City & Community*, 9(2), 145–64.

Yalch, R. and Spangenberg, E. 1990. Effects of store music on shopping behavior. *Journal of Services Marketing*, 4(1), 31–39.

Yang, W. and Kang, J. 2005. Acoustic comfort evaluation in urban open public spaces. *Applied Acoustics*, 66(2), 211–29.

Young, G., Bancroft, J., and Sanderson, M. 1993. Musi-tecture: Seeking useful correlations between music and architecture. *Leonardo Music Journal*, 3, 39–43.

Yu, L. and Kang, J. 2008. Effects of social, demographical and behavioral factors on the sound level evaluation in urban open spaces. *Journal of the Acoustical Society of America*, 123(2), 772–83.

Yu, L. and Kang, J. 2009. Modeling subjective evaluation of soundscape quality in urban open spaces: An artificial neural network approach. *Journal of the Acoustical Society of America*, 126(3), 1163–74.

Yu, L. and Kang, J. 2010. Factors influencing the sound preference in urban open spaces. *Applied Acoustics*, 71(7), 622–33.

Zanetti, F. 2012. Recorded interview with Author, 25 June 2012.

Zembylas, M. and Michaelides, P. 2004. The sound of silence in pedagogy. *Educational Theory*, 54(2), 193–210.

Zimmermann, S. 2007. Media geographies: Always part of the game. *Aether*, 1, 59–62.

Ziv, N., Sagi, G., and Basserman, K. 2008. The effect of looks and musical preference on trait inference. *Psychology of Music*, 36(4), 463–76.

Zonn, L. 2007. Going to the movies: The filmic site as geographic endeavor. *Aether*, 1, 63–7.

Index

Note: Figures and tables indexed in **bold**.

For Product Safety Concerns and Information please contact our EU representative GPSR@taylorandfrancis.com
Taylor & Francis Verlag GmbH, Kaufingerstraße 24, 80331 München, Germany

www.ingramcontent.com/pod-product-compliance
Ingram Content Group UK Ltd.
Pitfield, Milton Keynes, MK11 3LW, UK
UKHW010957180926
475701UK00004B/419

* 9 7 8 1 1 3 8 2 7 0 1 4 5 *